D·I·C·T·I·O·N·A·R·Y · O·F
INDUSTRIAL
SECURITY

S·T·E·W·A·R·T · K·I·D·D

D·I·C·T·I·O·N·A·R·Y·OF
INDUSTRIAL
SECURITY

S·T·E·W·A·R·T·K·I·D·D

Routledge & Kegan Paul
London and New York

First published in 1987 by
Routledge & Kegan Paul Limited
11 New Fetter Lane, London EC4P 4EE

Published in the USA by
Routledge & Kegan Paul Inc.
in association with Methuen Inc.
29 West 35th Street, New York, NY 10001

Set in 9/10pt Times
by Input Typesetting Ltd, London
and printed in Great Britain
by TJ Press (Padstow) Ltd, Padstow, Cornwall

© Stewart Kidd 1987

Library in Congress Cataloging in Publication Data

Kidd, Stewart, 1947–
    Dictionary of industrial security.

    Bibliography: p.
    1. Industry—Security measures—Dictionaries.
I. Title.
HV8290.K48   1987   658.4'7'0321   87–4672

British Library CIP Data also available

ISBN 0–7102–0794–8

To Chris, who never stopped believing that it would be finished, and only occasionally required reassurance.

CONTENTS

TABLES AND DIAGRAMS

INTRODUCTION

The growth of industrial security since the Second World War has to be one of the major commercial phenomena of the twentieth century, yet it has been little remarked on outside the ranks of those involved in the field. In the UK for example, before 1939, there was only one company providing security services. This company, the forerunner of the giant Securicor Group, provided fifteen bicycle-riding watchmen who, armed with truncheons and whistles, provided a patrol service for subscribers, mainly in London's Mayfair and Belgravia areas. The company had been set up to meet a need from householders (and insurance companies) who were concerned at the increase in burglaries of large houses.

In the US, private security actually substantially predates the establishment of public law-enforcement agencies. The establishment by the overland freight companies and later by the railroads of police departments (which incorporated investigative functions) anticipated the growth of organised police forces. It is worthy of note that when President Lincoln was in need of investigative assistance, it was to the Pinkerton Agency that he turned.

The security industry in the US is growing at a substantial rate, as a 1985 study shows. Produced by a well-known Wall Street company, the study estimates that security products worth more than US$2.4 billion were shipped in the fiscal year 1984–5 and predicts that this figure would increase to US$4.4 billion by the end of 1988, and US$8 billion by the mid-1990s. These estimates must in mid-1986 be seen to be conservative. Another study, published in March 1986, suggests that the total US domestic market for security products (excluding security services) in 1986 will be in excess of US$4 billion.

The reason for this growth, which is paralleled in Europe, is complex and relates not only to major increases in crimes against property but to changing social attitudes and a decline in traditional standards. The economic restructuring of the West, the 'post-Industrial Revolution' and even the microchip should all share part of the blame, as must the declining influence of church and school.

These changes in community standards as well as economic activity have been mirrored by a change in public policing, to the extent that many jobs once seen as purely police functions are now no longer part of the responsibilities of law-enforcement departments. Private industry has not been slow to fill this vacuum. Beginning in the 1950s, security companies expanded into the cash transport business. Then, in the 1960s, just as policemen started to leave footbeats for Panda (small patrol) cars and abandoned 'shaking hands with doorknobs', there was a dramatic increase in demand from industry for the provision of security guards. In the 1970s, and to a greater extent in the first years of this decade, we have seen private security companies taking over the monitoring of intruder alarms and providing what are essentially private police services within large shopping and residential complexes. The alarm monitoring function, in some parts of the US at least, requires that security companies using armed guards actually make the first response to an alarm call.

The consequences of these changes are reflected in the fact that in both the US and UK, there are now more persons employed in the private security industry than in law enforcement.

This then, is the background to one of the most important growth areas in the commercial sector (and one of the few industries which, like crime, is actually expanding). Security (meaning industrial and commercial security) can be said to have entered the first stage of becoming a profession. One of the most admired practitioners of this art, the late and much-missed Alex Smart once remarked that 'security would never become a profession until it developed an academic base' and he likened our position to that of the barber-surgeons of the early seventeenth century. Group-Captain Smart maintained that until the surgeons had developed a scientific basis for their skills and set up schools where the various aspects of their craft could be taught in an organised and rational manner, they were not accepted as a profession. It is hoped that it is towards this goal that developments in security education, especially in the US, are moving.

One of the criteria that set professions apart is the development of their own jargon. It would appear that some groups try intentionally to develop an argot which is incomprehensible to outsiders. While it is to be hoped that this is not the case for security, like the other professions, we have developed our own jargon, much of it based on terms used in law enforcement and other, longer-established professions. This wide borrowing reflects the range of jobs which the security professional is called upon to undertake. Perhaps most significant is the number of words drawn from technology, reflecting developments which in the past ten years have radically altered the ways in which the security professional can approach the problems of protecting people and property. The speed of change can best be reflected in that only a few weeks before the text of this book was sent to the printer, changes had to be made as a result of new equipment seen at a major exhibition.

It is inevitable that in a work of this size, the principal problem has been what to leave out rather than what to include and so a decision had to be taken to discard many words and terms which could reasonably be expected to be found elsewhere. This means that some terms which could have been included in this book have been omitted because they will be more conveniently found in for example, a good legal dictionary, a dictionary of fire technology or a dictionary of computer terminology. Those practising security in a field which involves substantial contact with architects or builders will also find access to a glossary of civil engineering or construction terms useful.

One of the intentions of the author was to produce a work which will not only be of use to the security professional but also to the non-specialist. For it is the manager who may have a supplementary responsibility for security or loss prevention who most desperately needs a guide through the morass of alarm terminology and the labyrinth of keys and locks.

This dictionary is also intended to be of use to those who are studying for one or other of the professional examinations in security, and it is hoped that the author's experience in preparing candidates (especially those whose first language is not English) for such examinations will not result in the criticism that some of the terms included in the work are 'obvious'.

The author would be pleased to receive comments from readers (including suggestions for the inclusion of additional words) for future editions. While grateful acknowledgment is made to those who have helped with the compilation of this work, the credit or blame for what lies before you is the author's alone.

Lock Terminology

One of the most confusing areas of security terminology is that part which relates to locks and locking devices. Some confusion has arisen in the past because of the fact that many locks and catches are referred to by more than one name and there appears to be no hard and fast rule as to which is 'correct' or appropriate. In this work, the most commonly used terms have been included and cross-referenced to similar terms. It should be noted that many lock manufacturing companies use their 'own' terms, and this does create confusion. The pre-eminence of one company in a particular field may create an inaccurate impression in the mind of the general public so that for example, many people talk about a 'yale lock' when they do not necessarily mean to specify a lock manufactured by Yale Industries, a subsidiary of the Eaton Corporation, who now own the company formed by Linus Yale. It is perhaps because the USA is the home of the pin tumbler lock that much of its development has taken place there. Accordingly, where there is a conflict in nomenclature, the author has adopted the alternative preferred by the ALOA for pin tumbler locks. At the same time, the UK lock companies have developed the lever lock to similarly high levels and so it would seem appropriate to use British terms where any conflict arises.

Security Standards

Appendix 1 contains a brief summary of British standards relating to industrial and domestic security, while Appendix 2 contains a slightly lengthier listing of some of the standards relevant to US security practice.

While there are those who decry standards, suggesting that they stultify technological development and are overused by the unimaginative, it is hard to see how technology could exist in its present form without the existence of consensus standards.

There are, of course, glaring gaps: there is no nationally-accepted standard in the UK for safes; Underwriter's Laboratories in the US have requested that a major security work does not make reference to the intruder detection equipment standards which do exist.

Standards are included in this work because of the author's experience in the developing world, where there is a dearth of standards, and where even when local standards exist, they are unlikely to include security matters. Easy access to standards numbers and titles is a boon to the security professional working outside the US and Europe, and it is hoped that those in the business who feel that'they have no need of standards will not feel too offended.

Security Bibliography

The bibliography contains a list of books on security topics. It does not pretend to be authoritative but an attempt has been made to cover all the major security specialities (aviation, hotels, hospitals and retail), as well as more technical matters like closed-circuit television and alarm systems.

There are a few books which should form a part of any security professional's library and these are marked with an asterisk*. These works will form the basis for self-study or correspondence courses, and possession of them will prove invaluable for the security examinations administered by ASIS and IPSA.

No mention has been made in the book list of non-security works, but those who are to practise in the UK will find Butterworth's reincarnation of *Moriarty's Police Law* invaluable.

Where a work is known to have been published in both the US and UK this is stated. Most of Butterworth's list is available in the UK, even though this is not specifically stated.

ACKNOWLEDGMENTS

I gratefully acknowledge all those who have helped in the production of this book and wish particularly to credit the assistance given by Mike Clewes FIISec, who has been a friend and collaborator on many projects, and by Basil Shannon, Research Officer, Royal Hong Kong Police Force.

The input from both these security professionals was invaluable, but particularly in respect of lock technology (an area which the author includes among his weaknesses), Basil Shannon's valuable and patient assistance is gratefully acknowledged.

Other thanks are due to Jamil Al Alawi, Undersecretary for Power and Water in the Ministry of Works Power and Water, State of Bahrain, for his support and understanding; to Warren Metzner CPP for revealing to me just what industrial security is all about; to Peter Hamilton MBE for his guidance and encouragement when I first started writing, and to Jim Cretecos CPP for his tolerant and patient tutoring in structuring a safety and security department.

The diagrams included in this work were originally prepared by the Audiovisual Aids Unit of the Training and Development Department, Power and Water Affairs, Bahrain, to whom I extend my appreciation and gratitude. I would also like to thank J. E. Reynolds & Co, manufacturers of the ERA range of locks and locking devices for permission to develop some of the drawings from their product range. Other drawings were produced from ideas by Basil Shannon.

Finally, I wish to mention Lt.-Col. Frank Collard who first inspired me to write and who patiently sought to impress upon me the correct use of English as a means of communication. Any failure in this respect is not his fault.

A

ABC Alarms By Carrier a system for transmitting both fire and security ALARMS using existing telephone lines. The system enables up to four different types of alarms to be transmitted on a single telephone circuit without interfering with speech messages. Similar systems are known as BASE 10, Poll Start and Red Care. Red Care is now being actively marketed by British Telecom. Poll Start is being developed by Bell Telephones in the USA.

ABI Association of British Investigators.

ABIS Association of Burglary Insurance Surveyors (UK).

access card (usually) a plastic card encoded mechanically or electronically, which when inserted into a CARD READER allows entry if the encoded information is valid. Information which can be encoded includes authorisation codes, FACILITY CODES and information regarding the time frames within which the card is valid. This is known as the time zone. Access cards are also sometimes known as coded badges or coded cards. See also CARD ACCESS SYSTEM.

access codes a sequence of numbers or letters, which if correct, allows entry into an area or building. Access codes are often entered by using a KEYPAD and/or an ACCESS CARD.

access control the use of manpower and procedures as well as electrical and mechanical components to regulate entry into specified locations.

access control system an interconnected arrangement of various types of equipment intended to control entry into a location.

access mode the state of an ACCESS CONTROL SYSTEM when entry to a location or piece of equipment is permitted.

access route normally the usual entry and exit route into or out of premises protected by an ALARM SYSTEM. The term is also occasionally used to refer to routes used when carrying cash.

acknowledgment a signal or command forming part of a pre-set procedure in which an operator accepts or acknowledges an ALARM or other incoming indication within a predetermined time.

AC line carrier a means of transmitting alarm signals via mains power lines.

1

acoustic detector see AUDIO DETECTION.

active card see HANDS-FREE ACCESS CONTROL.

active door an installation containing two doors has one active door, which must be opened first. See also LEAF.

active infrared (sensor) an INFRARED SENSOR which creates a detection beam or field by emission of infrared beams. Compare with PASSIVE INFRARED SENSOR.

active mode a state in which a SENSOR or system is operational.

active sensor a SENSOR which generates a field or beam or which radiates energy in order to protect a given location. See also PASSIVE SENSOR.

actuator an electromechanical device used either to initiate an alarm signal by converting physical activity into an electrical signal (e.g. a PRESSURE PAD), or to convert an electrical signal into mechanical activity (e.g. an ELECTRIC DOOR STRIKE).

aiming a 'trial and error' technique used to implement pre-planned lighting schemes by means of manually adjusting or moving light units to ensure that areas actually are lit in accordance with the specifications.

airlock see INTERLOCKED DOORS.

AIRMIC Association of Insurance and Risk Managers in Industry and Commerce. See also RISK MANAGEMENT.

alarm 1. a common name for a complete INTRUDER ALARM SYSTEM as in the phrase BURGLAR ALARM
2. the signal produced by such a system.

alarm bypass a feature of intruder alarms which permits a KEYHOLDER or other authorised person to enter alarmed premises without triggering the alarm. The device works by SHUNTING specific components of the system such as door contacts out of the ACTIVE MODE, starting a time sequence within which the alarm system must be switched off. Failure to switch off the system within the pre-set time will cause the alarm to be triggered.

alarm condition the state of any part of an INTRUDER ALARM SYSTEM which indicates that an intrusion (or attempted intrusion) has occurred or is about to occur. When a system is registering an alarm condition, it is said to be 'in alarm'.

alarm control (panel) the heart of any alarm system is the CONTROL PANEL which switches the system on and off, receives signals from detection devices and sends out signals to peripheral components such as bells, DIALLERS or to a CENTRAL STATION. Control panels also usually house other components such as batteries, chargers and some form of mechanism to SET and UNSET the system such as a KEY SWITCH.

A

alarm discriminator a device intended to reduce or eliminate the incidence of unwanted or spurious alarms. Such devices normally form part of the circuitry of the system's CONTROL EQUIPMENT (or may be part of the SENSOR's circuitry), and work by electronically filtering incoming signals and rejecting those which do not meet pre-set criteria. An example of this would be found in a system incorporating VOLUMETRIC DETECTION which was programmed to reject alarms created by birds or small mammals.

alarmed exit device see CRASH BAR.

alarm indicating device any audible or visual signal intended to advise of the presence of an unusual situation.

alarm initiating device a detection device or sensor used to trigger an alarm either automatically, or, in the case of BANDIT ALARMS, manually.

alarm mode see ALARM CONDITION.

alarm output any signal produced by a CONTROL PANEL intended to warn audibly or visually of the existence of an ALARM CONDITION. In particular, alarm outputs normally activate LOCAL or EXTERNAL AUDIBLE ALARMS or SOUNDERS.

alarm receiver that part of an alarm system's circuitry which registers that an alarm has been triggered. This is normally indicated both audibly and visually on the CONTROL PANEL by a light-emitting diode (LED) and a buzzer.

alarm reporting system the means by which an ALARM is transmitted from the control panel to monitoring personnel.

alarm signal an electrical, optical or electromagnetic signal (which may be audible or visual) indicating that an ALARM has been triggered.

alarm station a manually operated device, usually in the form of a switch or button which will trigger an alarm. These devices may be in the form of foot-operated switches and are used extensively in banks and similar locations. An essential element of such devices is that they are always operable, being connected to a PERMANENT CIRCUIT. These devices may be known by a number of names – attack buttons, bandit alarms, hold-up alarms, panic alarms, personal attack buttons, pull buttons or raid alarms.

alarm system a collection of SENSORS, signalling devices and CONTROL EQUIPMENT which forms the basis for a system which will detect the presence of intruders, fires or other unusual occurrences.

alarm transmission the signal sent from a CONTROL PANEL to a CENTRAL STATION.

alarm transmission circuit the route by which an alarm is sent from the CONTROL PANEL to the CENTRAL STATION or other monitoring point. See also DEDICATED LINE.

ALOA Associated Locksmiths of America, a professional association with many international members which conducts a proficiency registration program for locksmiths. Grades are RL – Registered Locksmith; CPL – Certified Professional Locksmith, and CML – Certified Master Locksmith.

ALSIA Association of Licensed Security and Investigative Agencies (South East Asia).

ambient light the level of light normally found in a given area. The light may be either daylight or artificial light.

ambient temperature the temperature range normally found in a given location.

ambush code a code usually associated with ACCESS CONTROL SYSTEMS. Use of this code will trigger an alarm, usually at a remote location to indicate that the person inputting the ambush code is acting under duress or is being held hostage. For example, one could program(me) an access control system to initiate an alarm if, instead of three digits being entered into a keypad, five digits were entered. The access control system would still operate normally and the intruder or hostage-taker would not be aware that an alarm had been triggered.

angle of view the field of vision of a closed-circuit television or film camera. See also FOV.

annealed glass 'ordinary' (float, plate or sheet) glass which has little resistance to impact or splintering. Rough-cast glass (which is also produced by rolling) has similarly little security value.

annunciator a device which indicates the existence of an ALARM CONDITION. Annunciators are normally audible, visible, or both.

anti-climb paint a specially formulated paint which can be applied to poles, drainpipes and the like to deter or delay intruders. Such paint never dries and will stick to the hands and clothes. The presence of the paint could provide useful evidence for later detection and prosecution, particularly if it contains a CHEMICAL TAG.

anti-cutting roller see ROLLER.

anti-eavesdropping device an electronic device resembling a radio receiver which automatically scans transmission frequencies to detect and indicate the presence of a BUG using radio transmission. See also SECURITY SWEEP.

anti-passback circuit a facility found in programmable ACCESS CONTROL SYSTEMS which ensures that a card user must enter and then leave an area before access is permitted a second time. This prevents the card being passed to an unauthorised person e.g. over a fence. See also PASSBACK.

anti-scaling barrier the use of BARBED WIRE or other materials, including iron spikes and PALISADE FENCES, to create barriers around possible access points such as flat roofs, drainpipes and balconies.

anti-shim device a feature of certain types of SPRING LOCK which prevents the springbolt from being depressed by LOIDING (or shimming) when the door is closed. A TRIGGER is also fitted to some latches which, when activated, DEADLOCKS them against loiding. See also ANTI-THRUST SLIDE.

anti-tamper contact the active component on an ANTI-TAMPER DEVICE, normally a microswitch.

anti-tamper device a CLOSED-CIRCUIT device designed to generate an ALARM CONDITION in response to any interference or attempt to TAMPER with any component part of an alarm system (providing such components are so protected). See also TAMPER SWITCH.

anti-thrust slide an auxiliary BOLT operating alongside and in contact with a LATCH BOLT, which, when depressed by the striking plate acts as an ANTI-SHIM DEVICE DEADLOCKING the latch.

Anton Piller Order an English High Court order which entitles the person to whom it is issued to enter named premises and seize documents or other articles. The order is frequently used in cases involving counterfeiting of products.

APA American Polygraph Association. See alo POLYGRAPH.

aperture a device which controls the amount of light entering a camera. The value of the aperture is expressed as an 'f' number which also indicates the speed of the lens system and its ability to gather light. See also AUTOMATIC IRIS, IRIS.

APTS Association for the Prevention of Theft in Shops, a UK-based association of major retailers and other interested parties who promote programmes to deter shop theft.

area detection see SPACE PROTECTION.

area lighting general illumination (usually by floodlighting) of a particular area for security, safety or night working.

area protection see SPACE PROTECTION.

armoured front see CYLINDER GUARD.

armoured glass a combination of glass, plastic or polycarbonate sheets of varying thicknesses used to protect cashiers, counter surfaces and shop windows. See also BULLET-RESISTANT GLAZING; BANDIT-RESISTANT GLAZING.

armoured pin the first PIN in a high security PIN TUMBLER MECHANISM.

The pin is made of hardened steel (instead of brass or nickel) so as to resist drilling or MANIPULATION or PICKING.

armoured shutter a recent development in banking security permits the installation of a bullet-resistant, armoured shutter which is usually recessed into the counter. Powered by a hydraulic motor or compressed air the shutter can be raised in less than a second, completely protecting the bank staff. The shutter mechanism can be linked to a BANDIT ALARM or SUSPICION BUTTON.

armoured vehicle 1. a vehicle constructed for the carriage of cash or other valuable cargoes, normally on a converted commercial chassis **2.** a specially converted saloon or limousine used for the transportation of those who are likely terrorist or kidnap targets.

ASET Academy of Security Educators and Trainers (US).

ASIS American Society for Industrial Security (International) US-based professional association with extensive worldwide membership at security manager level. See also CPP.

ASSE American Society of Safety Engineers.

assets the various parts which make up a commercial undertaking. The term is used in security contexts to include everything of value both tangible (plant, premises) and intangible (patents, expertise, staff loyalty and reputation). 'Assets protection' is a US term which is increasingly in use as a synonym for 'INDUSTRIAL SECURITY'.

assets protection see ASSETS.

ASTM American Society for Testing and Materials, a US-based body which produces a number of consensus standards in industry.

ATI Armored Transportation Institute, US-based association of companies operating armoured cash-carrying vehicles.

ATM Automatic or Automated Teller Machine, a microprocessor-controlled machine which gives access to a number of banking services including the dispensing of cash. Automated teller machines must be provided with adequate physical and electronic security. Access to services is normally obtained by inserting a coded MAGNETIC CARD and inputting a PIN through a KEYPAD. Failure to input a pin which relates to the encoded data from the card results in denial of access and retention of the card by the machine.

attack button a manually-operated alarm SENSOR, connected to a PERMANENT CIRCUIT, normally with key RESET, used to trigger an alarm. See also ALARM STATION.

attractive items a term used in retail and industrial security for small or very expensive items which are highly likely to be stolen. In retail terms, such items would include watches, jewellery and furs. In industry,

A

the term would include test meters, tools, electrical spares and precious metals.

audible alarm an ALARM SYSTEM which indicates a breach of security only by means of a SOUNDER. The term can also be used to apply to a component part of a larger system where the audible alarm is intended to draw the attention of the operator to a particular part of the system or a change in MODE.

audio detection a method of detecting an intrusion using sound. Such systems use microphones and are installed in vaults or use SEISMIC DETECTION or GEOPHONES to detect tunnelling or intrusion at a perimeter.

audio monitoring a technique utilised in high security locations using microphones installed in vaults or similar facilities. To determine whether or not an intrusion is taking place, an operator can 'dial-in' to the AUDIO DETECTION SYSTEM and actually listen to any activity in the PROTECTED AREA.

audit roll the paper roll installed in many cash registers upon which is printed details of transactions together with a serial number. The audit roll is still found in electronic cash registers although it is becoming less frequently seen in stores where there are on-line cash terminals. The audit roll is a useful method of verifying sales, payouts and 'no sale' transactions and should not be accessible to cashiers. See also TEST PURCHASE.

authorisation level 1. a security rating or clearance to hold documents **2.** an approval to gain access to restricted areas or equipment **3.** specific permission to approve payments or cash transactions to a prearranged limit. In retail security, the term is often used with respect to the authorisation of credit card transactions which exceed the FLOOR LIMIT.

authorised access control switch a US term for a device used to SET or UNSET a system. Also known as a day/night switch or SUBSCRIBER UNIT.

authorised access switch another term for an ALARM BYPASS or SHUNT SWITCH.

auto-alarming switcher in closed-circuit television security systems, a control device which automatically selects and calls up a CAMERA view of an area in which an INTRUDER ALARM sensor has been triggered.

automatic deadlatch a LATCH in which the main BOLT is automatically DEADLOCKED when the door is shut.

automatic dialling equipment see TAPE DIALLER.

automatic iris a component part of a closed-circuit television CAMERA which provides automatic adjustment to the IRIS (which controls the APERTURE) to compensate for changing light levels. The inclusion of such a component not only ensures better quality images on the monitors, but also prevents damage to the camera components as a result of excessive light levels.

automatic McCulloh (system) see McCULLOH-TYPE SYSTEM.

automatic mortice locking latch see AUTOMATIC DEADLATCH.

background screening a civilian term for SECURITY CLEARANCE or vetting. Screening is an essential part of any company security programme and checks should be made not only on claimed work history but also on educational and professional qualifications. The more responsible the post, the more detailed the check must be. See also BONDING, NV, PV.

back plate the plate on the inside of a door which enables a PIN TUMBLER mechanism to be clamped to the door using the cylinder-fixing screws.

backup battery a source of reserve power used by an electronic or electrical system in the event of AC mains failure. See also UPS.

badge in security, this term is usually synonymous with IDENTITY CARD (especially when the card is worn on the person). Badges are normally made of a LAMINATED sandwich of paper and plastic and serve both as a means of personal identification and in ACCESS CONTROL. A badge is often combined with an ACCESS CARD which may be encoded either magnetically, mechanically or electronically. In these cases they are sometimes known as coded badges. Recent developments in printing technology have led to the availability of a self-destructing paper badge. This can be issued to visitors in the form of a self-adhesive label, and exposure to light during the course of a day will darken it, making it illegible and therefore incapable of being re-used. See also CARD ACCESS SYSTEM.

badge reader a device, forming part of an ACCESS CONTROL SYSTEM designed to read and interpret the encoded information on an ACCESS CARD or BADGE. See also CARD READER.

bait money bank notes intended to assist the police in tracing thieves. Bait money is normally marked using an INVISIBLE STAINING COMPOUND or the serial numbers of each bank note are recorded. Some cash drawers are fitted with a special compartment or clip for bait money. The clip is connected to the intruder alarm system. If the bait money is removed from either location, this triggers a BANDIT ALARM.

balanced hydraulic detection system see BALANCED PRESSURE SENSOR; PRESSURE DIFFERENTIATING SYSTEM.

balanced magnetic switch see MAGNETIC CONTACT.

balanced pressure sensor this device forms part of a SEISMIC DETECTION system and comprises two buried tubes each filled with a hydraulic fluid. When an intruder stands on the ground above the tubes, the weight of the intruder creates a pressure differential in the tubes which is sensed by a series of transducers. Any change beyond predetermined values is noted by the control circuitry which then triggers the alarm. See also BURIED LINE INTRUSION DETECTOR.

bandit alarm a manually-operated sensor forming part of an alarm system which when operated, triggers an alarm or operates BANK CAMERAS. Bandit alarms are always connected to a PERMANENT CIRCUIT. See also ALARM STATION.

bandit-resistant glazing a type of glazing intended to provide protection for staff handling cash or other valuables. Normally consists of a sandwich of glass and polycarbonate material at least 9.5 mm thick in a glazed opening not exceeding 2 m square. The supply and installation of this type of material should be handled by specialist contractors and care should be taken to ensure that an appropriate standard is followed. See also BULLET-RESISTANT GLAZING.

bank camera a CAMERA which uses standard 16 mm or 35 mm cine film to provide a permanent record of an incident. Such cameras are arranged to provide maximum SURVEILLANCE of the public areas and are connected into the BANDIT or SUSPICION ALARM circuitry of an alarm system so that if the system is triggered the cameras start automatically. The frames are normally exposed at a much slower rate than in a cine camera and the photographs produced often provide useful evidence for later arrest and conviction. In the US, some banks use these cameras to take photographs of individuals carrying out unusual or high-value transactions. In some cases, individual cameras can be triggered by cashiers who are suspicious of an individual. These cameras are also known as 'demand cameras', 'hold-up cameras' or 'suspicion cameras'. Recently, as a result of an increasing number of ATM frauds, bank cameras and (in some locations) closed-circuit television cameras with recording equipment are being installed inside ATM machines.

Bank Protection Act US legislation which mandates minimum security requirements for specified banking premises.

Bank Secrecy Act US legislation intended (among other things) to prevent MONEY LAUNDERING. Also known as Title 31.

Banshee see YODELARM.

barbed tape an improved type of barbed wire, originally developed in West Germany in 1952. It can be used as a direct replacement for BARBED WIRE and being manufactured from galvanised steel strip is wider and flatter than barbed wire. Barbed tape is also produced in a concertina format. The US now manufactures a much modified form of barbed tape to a higher specification and this product is known as GPBTO – General Purpose Barbed Tape Obstacle.

barbed wire galvanised wire manufactured with a small wire barb at regular intervals (usually about 300 mm). The wire is used to protect the tops of fences but it offers little resistance to intrusion and should be seen only as a deterrent against casual trespassers. Coils of barbed wire (sometimes known as Dannert or concertina wire) are used extensively in military applications.

B

barrel bolt a surface-mounted, non-locking BOLT where the 'round' or bolt SHOOTS in a continuous guide or barrel.

barrel distortion the distortion of an image on a television MONITOR which gives the picture a convex image.

barrel key a key with circular cross-section, with a hole in the centre of the SHANK or stem. A US term for PIPE KEY.

base 10 an ABC system of alarm transmission originally developed in Canada. Similar systems are being offered under the name of Poll Start (in the US) and Red Care (in the UK).

BDA British Detectives Association.

beam in the security context, normally taken to be a flow of light or propagation of an electromagnetic field in one direction within a confined path.

beam angle the angle between the outer limits of a BEAM.

beam interruption detector a type of detection device based on the INFRA-RED BEAM which triggers an alarm if the beam is broken.

beam splitter an optical device used to split a BEAM into a number of subsidiary beams to provide more effective coverage. Used, for example, in an INFRARED BEAM system.

bell delay a device forming part of the circuitry of an alarm system which delays activating the SOUNDER for a predetermined period. This delay is often found in alarm systems which are connected to CENTRAL STATIONS and is intended to allow the police to arrive before an intruder is aware that he has triggered an alarm.

bell shutoff a circuit within the CONTROL EQUIPMENT of an alarm system which switches off a SOUNDER after allowing it to operate for a pre-set period. In the UK, most commercially-installed INTRUDER ALARM SYSTEMS are fitted with such a circuit, which shuts off all sounders after 20 minutes. This feature is sometimes known as 'West Midlands Policy' after the police force which originally mandated a bell-shutoff requirement.

bicentric pin tumbler cylinder a CYLINDER LOCK which has two PLUGS and two sets of PINS which requires two different keys to operate it. Mainly used to obtain a higher level of security in MASTER KEYED SYSTEMS.

bill trap a slot or carousel fitted to a cashier's window to allow bank notes or documents to pass from one side to the other.

11

biometric sensor a method of identification used in ACCESS CONTROL SYSTEMS which operates by measuring a particular feature of the human body. Such systems include equipment based on the measurement of finger and palmprints, retinal patterns, signatures, and voice patterns. All of these systems are commercially available and offer perhaps the most significant advances in security technology since the invention of the microchip. While the SIGNATURE VERIFICATION SYSTEM, which operates by means of a variety of pressure sensors programmed to record the pressure and angle of the pen and duration of the signature offer many advantages for the financial sector, it would seem likely that either the fingerprint system or the retinal scan are likely to take the major share of the high security ACCESS CONTROL market. See also EYE-DENTIFY, FINGER-MATRIX, IDENTIMAT.

bit in security terminology, the portion of a KEY (usually for use in a WARDED or LEVER LOCK) which protrudes and engages with the bolt and/or the levers of the lock. See also BLADE. (See diagrams 5 and 6 on p. 61)

bit key a key with a bit projecting from a round SHANK, similar to the BARREL KEY, but of solid construction. Examples are LEVER LOCK KEYS and PIPE (or barrel) keys.

bitt see BIT.

bitting 1. the actual cuts in a key (or key cuts) forming the COMBINATION. Also known in LEVER LOCK KEYS as lever notching or in CYLINDER KEYS as notching
2. numbers representing the dimensions of key cuts. (See diagrams 5 and 6 on p. 61)

black body radiation the small amount of heat emitted by objects or people which makes them stand out from the background, thus making them visible on the monitor of a closed-circuit television system using PASSIVE INFRA-RED.

black box see PHONE FREAK.

blade the part of a key, usually a PARACENTRIC key, on which the COMBINATION or BITTING is cut. The term 'bit' is generally used for LEVER LOCK KEYS and 'blade' for the 'long bit' of PIN TUMBLER LOCK keys. (See diagram 5 on p. 61)

blank 1. an unfinished or uncut key
2. material (of any type) which is of the correct size and shape to be allowed to enter the KEYWAY of a lock (but without the combination cuts).

blind area 1. in intruder alarm systems, an area that is not covered by a SENSOR.
2. in closed-circuit television systems, an area not covered by a camera.
3. an area in which a radio set can neither receive nor transmit.
Also known as a blind spot.

B

blind spot see BLIND AREA.

BNC connector a COAXIAL CABLE connector using a bayonet-type plug and socket system, commonly used in closed-circuit television systems.

bolt 1. a metal bar of either flat or round cross-section used to secure a door or window by means of the head of the bolt slotting into a counter-sunk hole on the frame. This bar is also known as a round
2. the part of a lock which moves in and out of the bolt hole in the STRIKING PLATE when the key is turned. (See diagram 7 on p. 67)

bolt stump the part of a DEADBOLT which passes through a slot or GATE in the levers as the bolt moves. (See diagram 7 on p. 67)

bolt throw the distance a BOLT travels when operated, usually referring to a DEADBOLT. See also LOCK SHOOT.

bomb blanket a device specially designed and constructed to suppress and contain the explosive force of an IED.

bomb disposal the term commonly used to describe the neutralisation of unexploded ordnance or similar devices by specially trained personnel. Other terms used for this work include 'explosive ordnance disposal' and 'ballistic disposal'.

bonding a form of insurance against the loss of cash resulting from theft or fraud by employees. This insurance is also known as fidelity bonding. Cover is obtained only after BACKGROUND SCREENING is carried out.

booster bag a slang term for a large bag (usually with hidden pockets) used by shoplifters. See also SHOPLIFTING.

booster box a slang term for a SHOPLIFTING aid normally consisting of a large empty carton, wrapped to resemble a parcel and having a flap cut in one side or in the base. Items can be pushed in through the side flap or the box can be placed on a counter over small items to trap them.

bottom pins see PIN TUMBLER MECHANISM. (See diagram 10 on p. 70)

boundary fence see FENCE.

bow the head or handle of a key. (See diagrams 5 and 6 on p. 61)

box a metal closure forming part of a STRIKING PLATE which protects a DEADBOLT. See also BOX STRIKING PLATE. (See diagram 9 on p. 69)

box lock a term occasionally used (incorrectly) for RIM LOCK.

box staple a surface-mounted BOX STRIKE. (See diagram 12 on p. 71)

box strike see BOX STRIKING PLATE.

box striking plate a STRIKING PLATE which provides a fully enclosed housing to receive the lock BOLT, offering better protection (for the bolt)

B

in case of an attack which places end-pressure against the bolt with a JEMMY or similar tool. (See diagram 12 on p. 71)

break alarm 1. an ALARM SIGNAL produced by opening or breaking an electrical circuit. Also known as an OPEN-CIRCUIT ALARM 2. an alarm SENSOR comprising a closed switch fitted to a glass plate. If the plate breaks, the switch opens, triggering an alarm. Now fairly uncommon.

break-beam alarm an alarm system or SENSOR which is triggered by any interruption of a beam of visible or invisible light. See also INFRA-RED BEAM, LASER PROTECTION SYSTEM, PHOTOELECTRIC SENSOR.

BS British Standard (Specification), a specification or standard issued by the British Standards Institution. See Appendix I for a list of standards relevant to the security profession.

BSIA British Security Industry Association, a trade body whose membership comprises the principal companies supplying security services and to a lesser extent, equipment. The Association has an Inspectorate which is intended to ensure that member companies maintain standards of service.

bug 1. a covert sensor or listening device 2. to introduce or plant such a device.

bullet cuts see BULLET WARDS.

bullet-resistant glazing BS 5051 lists 5 grades of such glazing. In each case, the glazing under test must resist the impact of the projectiles without their penetrating:

G0 – 3 rounds fired at 3 m from 9 mm military parabellum handgun
G1 – 3 rounds fired at 3 m from .357 magnum handgun
G2 – 3 rounds from 7.62 mm NATO rifle
GS – 2 shots fired at 3 m from full choke 12 bore shotgun
Anything less than this should not be classified as bullet-resistant.
UL Standards list similar criteria (see Appendix 2). See also BANDIT-RESISTANT GLAZING.

bullet wards 1. the projections or grooves in a lock KEYHOLE designed to prevent entry of an improperly-shaped key. (See diagram 9 on p. 69) 2. the longitudinal projections or grooves on a key BIT to permit the key to enter a bullet-warded lock. Sometimes known as bullet cuts. (See diagram 9 on p. 69)

burglar alarm the commonly used term for an INTRUDER ALARM SYSTEM.

buried cable system see BURIED LINE INTRUSION DETECTOR.

buried detection system the common term for any detection system which relies on SEISMIC DETECTION, PRESSURE DIFFERENTIATING, or MAGNETIC BURIED LINE SENSORS.

buried (line) intrusion detector a term normally applied to a PERIMETER PROTECTION SYSTEM using either electromagnetic radiation or the PRESSURE DIFFERENTIATING SYSTEM. Although other buried systems (using for example GEOPHONES) do exist, these are normally referred to as SEISMIC DETECTION SYSTEMS. See also BALANCED PRESSURE SENSOR, ELECTRIC FIELD SENSOR, MAGNETIC BURIED LINE SENSOR.

burn-in a term used to describe the 'running-in' of any electronic system and in particular, a closed-circuit television system. The process is often done at increased stress (higher voltages, light levels and input) prior to handover to ensure that any defective components reveal themselves.

B

C

cablelock a type of PADLOCK to which is permanently attached one end of a strong chain or steel braided cable. The free end of the cable fits into the padlock enabling the device to be used to secure items to fixed points – for example, through a bicycle frame and around a lamp post.

caltrop a device consisting of four sharp spikes, so arranged that however the device is placed on the ground one of the spikes is always vertical. Originally a military weapon used to disable horses, now used to prevent vehicles crashing road-blocks or police check points. A more effective method of achieving this is to use DRAGON'S TEETH. Caltrops are sometimes known as horse irons.

cam a flat tongue or actuator which is attached perpendicularly to the rear of a PLUG and which turns as the key is rotated. (See diagram 10 on p. 70)

cam catch a non-locking device often found fitted to SASH WINDOWS, also known as a fitch catch. See also LOCKING CAM CATCH.

camera in security terms 'camera' normally refers either to a closed-circuit television camera or to a surveillance camera. Closed-circuit television is extensively used in almost all types of industrial and commercial security but surveillance cameras (sometimes called BANK or HOLD-UP CAMERAS) are normally only found in banks, building societies and savings and loan offices or other similar institutions.

camera dome a method of mounting a closed-circuit television camera inside buildings. The dome is normally of one-way glass and fitted to a ceiling or roof so it is not possible to tell in what direction the camera is pointed. The dome also provides physical protection for the camera and lens.

camera housing a metallic casing in which cameras are installed to prevent either tampering or environmental damage. External housings are usually fitted with a variety of accessories including heaters, fans, demisters, wash and wipe units and ANTI-TAMPER DEVICES.

cam lock a cylindrical locking device whose CAM is the actual locking bolt.

16

capacitance the measure of or the ratio of the accumulated charge within a CAPACITOR to the applied voltage.

capacitance alarm (system) an alarm system normally used to protect safes or filing cabinets which operates by generating a field which effectively turns the whole safe or object being protected into an aerial or antenna. If any part of the protected object is touched, the capacitance changes, triggering an alarm. This type of system is also often used as a SAFE ALARM or proximity alarm.

capacitance coded card an ACCESS CARD using the capacitance principle, which offers excellent security but at a comparatively high cost. (See table 1 below)

capacitance sensor a type of sensor used in a CAPACITANCE ALARM SYSTEM.

capacitative alarm system another term for CAPACITANCE ALARM SYSTEM.

capacitor an electrical or electronic component capable of building up an electrical charge supplied to it by a source of current.

card access (system) an ACCESS CONTROL SYSTEM using one or other of the varying types of cards available. (See table 1 below)

Card	Cost	Storage capacity	Security against tampering	Subject to abrasion	Can be duplicated	Affected by temperature	Affected by magnetic fields
Magnetic strip	Low	High	No	Yes	Yes	Moderate	Yes
Magnetic spot	Moderate	Medium	Yes	Moderate	Yes	No	Yes
Weigand effect	Moderate	Medium-high	Yes	Yes	No	No	No
Infrared or optical density	Moderate to high	Medium-high	Yes	No	No	No	No
Capacitance	Moderate	Medium	Yes	No	No	No	No
Hollerith ('punch cards')	Low	Low	No	Yes	Yes	No	No

Table 1   *Comparison of access cards*

card encoder a device for electrically, magnetically or mechanically coding an ACCESS CARD.

card key an ACCESS CARD. Cardkey also is the proprietary name for the access control products of a major US security equipment manufacturer.

card reader a scanning device which interprets encoded information from an ACCESS CARD. Card readers may be STAND-ALONE, in which case, the information scanned will also be interpreted within the card reader's circuitry, or ON-LINE. In the latter case, the card reader will be permanently connected to a remote computer or control system.

carrier-current receiver a device used in the detection of covert or illicit listening devices which will indicate the presence of BUGS drawing current from appliances in a room and using the AC cables to transmit signals. Note that this device detects only bugs HARDWIRED into the building's AC mains supply. See also DEBUG; SECURITY SWEEP.

carrier-current system a type of wire-less ALARM SYSTEM using the existing AC power system within a building. Alarm components are plugged into the normal power outlets or sockets. It is important to ensure that all sockets used are wired on the same phase of the supply system or the alarm will not work. The system offers advantages of cost and speed of installation as the need to HARDWIRE components is largely eliminated.

carrier receiver a device which will monitor a wide range of radio frequencies to detect the presence of a carrier wave, which would indicate the presence of an illicit listening device.

cascaded intensifier see SIT.

cascade system a method of passing information rapidly to many locations. Often used in retail establishments to pass information regarding shoplifting gangs or credit card and cheque fraudsters. The system works when establishment A (which has the original information) telephones shops B, C and D and advises them of the problem. In a predetermined sequence, B, C and D call three shops in turn, each of those three calling three more. Thus, in under five minutes, up to forty locations can receive important information.

case the metal box into which are fitted the component parts of a lock. (See diagram 9 on p. 69)

casement a window, normally of the type which opens on vertical hinges.

casement fastener see COCKSPUR HANDLE.

casement lock any locking device intended to secure a CASEMENT, window.

casement stay screw a SUPPLEMENTARY LOCKING DEVICE which fits through the hole in a non-locking WINDOW STAY securing it to the frame. (See diagram 1 on p. 19)

casement stay stop a device similar to a CASEMENT STAY SCREW which clamps the stay arm to the window sill.

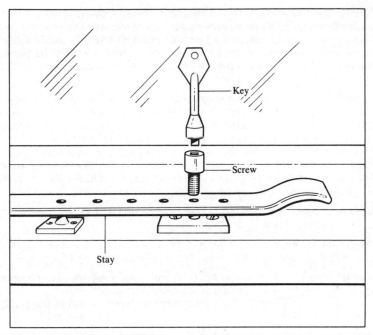

Diagram 1 *Casement stay screw*

cash box 1. a metal, lockable box normally used for the safe custody of small amounts of cash, stamps etc. Has no security value and should be secured in a SAFE or VAULT overnight
2. A large, sealable box, made of metal, fibre or polyethylene in which large amounts of cash, cheques or securities are transferred from place to place. Such boxes are sometimes called courier dispatch boxes.

cash-in-transit see CIT.

CATV Common Antenna TV (system) or Communal Aerial TV (system): a method of distributing broadcast television pictures throughout a large building. The system has a security implication because it is possible to use the distribution network to allow tenants to see visitors prior to admitting them via a DOOR VIEWER SYSTEM.

CCD Charged-Coupled Device: the circuitry within a relatively new type of closed-circuit LOW-LIGHT TELEVISION CAMERA. Such equipment is still being developed, and although cameras are now available which offer advantages of low weight and reduced power requirement as well as smaller size, these advantages have to be offset by relatively poor RESOLUTION and restriction on the use of some types of lenses.

C

CCTV Closed-Circuit TV (system): a method of providing remote surveillance of selected locations using small television cameras and cabling. The systems are called 'closed-circuit' because there is no broadcast signal. Increasing use is being made of means other than the traditional COAXIAL CABLE for signal transmission using, for example, FIBRE OPTICS or microwaves.

central station a specially designed and equipped location which monitors (on a 24-hour basis) both fire and INTRUDER ALARMS. Central stations may be operated by an alarm company or a CONTRACT SECURITY COMPANY or may be operated solely by a major company for its own use. Central stations, especially in the UK, are of increasing importance following the decision by the police to discontinue the monitoring of INTRUDER ALARM SYSTEMS. In addition to alarm monitoring, central stations will normally provide a KEY HOLDING service.

central station system an ALARM SYSTEM which is connected to and monitored by a CENTRAL STATION.

chain link fence a very common type of security fence constructed of interlinked galvanised steel mesh, frequently covered by PVC.

change key 1. a key which operates only one lock (or one group of locks) in a KEYED ALIKE SYSTEM. (See table 3 on p. 76)
2. a key used in some KEYLESS COMBINATION LOCKS to enable the combination to be altered.

change key lock a lock which can be operated by any one of a series of specially manufactured keys. A key is selected and the lock then set to operate only with that key. If, for example, the key is lost and it is necessary to change the combination of the lock to that of another key, this can be done simply. Change key locks are often used on safes and safety deposit boxes.

charged-coupled device see CCD.

checker an individual employed on a site or in an industrial facility to physically count incoming and outgoing merchandise or to weigh bulk deliveries and collections. Checkers are also sometimes used to supervise the CLOCKING-ON and CLOCKING-OFF of employees. Also known as a tally clerk.

chemical tag a method of property identification in the form of paint which includes a minute but uniquely identifiable chemical composition to enable a positive statement of ownership to be made. Similar chemical tags can be incorporated in ANTI-CLIMB PAINT to assist in the apprehension of intruders, and is also often included in commercial explosives.

cheque card a plastic EMBOSSED CARD, sometimes a MAGNETIC CARD, issued in the UK and in parts of Europe by financial institutions and used to guarantee a personal cheque providing that the retailer accepting the cheque follows certain rules.

cheval de frise a temporary barrier used to close roads, especially during emergencies. It is formed from either wood or angle iron and most closely resembles the military KNIFE REST but it is made substantially stronger to resist slow speed vehicle impact and is festooned with barbed wire or barbed tape. For greater effect, a cheval de frise should be chained at either end to a suitable fixed object and be backed up by a row of DRAGON'S TEETH. Where the device is being used semi-permanently as a vehicle barrier, it should have one third of its leg length buried and be constructed of material such as scaffolding pipe.

C

CHPA Certified Healthcare Protection Administrator: a designation awarded to hospital and healthcare security personnel after training, examination and career review by the IAHS.

CII Chartered Insurance Institute (UK).

CII Council of International Investigators (US).

circuit shunt a SHUNT SWITCH used to isolate all or part of a detection circuit from the CONTROL PANEL.

CIT cash-in-transit: the carriage of cash, bullion or other valuables in an ARMOURED VEHICLE or using other means of transport. Major financial institutions may have their own IN-HOUSE cash-in-transit unit.

CLASSIC Covert Local Area Sensor System for Intruder Classification: a proprietary name for a type of military RGS system.

clear zone the area either side of a fence or other PERIMETER BARRIER which should be kept clear of vegetation or rubbish.

clevis a metal link used to attach a chain to a PADLOCK.

clock see WATCHMAN'S CLOCK.

clock a UK police slang term (used in security). It refers to a specific observation made of a suspect or other person in whom there is an interest.

clocking fraud the offence of obtaining pay for hours not worked, usually by having an accomplice stamp a TIME CARD.

clock on/off the activity of coming on or off duty whether or not actually accompanied by the use of a TIMECLOCK and TIME CARD.

closed the state of an alarm system in which any of the detection devices or sensors can be triggered to produce an ALARM CONDITION.

closed-circuit an alarm system or component where all the sensing devices are wired in series and switches are in the closed position. Should a switch be opened or a wire cut, the alarm is triggered. This type of system is sometimes referred to as having continuous wiring.

closed-circuit television see CCTV.

closed detection circuit a detection circuit which is in an ALARM CONDITION when it is open.

close protection see EXECUTIVE PROTECTION.

close shackle padlock (sometimes, incorrectly written as closed shackle) a PADLOCK, usually of substantial construction, having the minimum amount of SHACKLE visible when in the locked position. This reduces the possibilities of cutting or forcing the shackle. Most LOCKING BARS specified by insurance companies are designed to be used with close shackle padlocks.

closing signal see OPENING SIGNAL.

CML Certified Master Locksmith, the highest grade of registration awarded by the ALOA.

coaxial cable a special shielded cable consisting of a central conductor (usually, a thick single copper strand) covered by a plastic sleeve. This insulator has a braided or woven copper wire shield over it and the cable is finished with a tough plastic outer cover. Coaxial cable is extensively used for closed-circuit television (see CCTV) and for aerial feeders for radio equipment.

cockspur handle a pivoting, non-security catch frequently found on older, metal-framed windows. Also known as a casement fastener.

cockspur handle stop a SUPPLEMENTARY LOCKING DEVICE used to secure COCKSPUR HANDLES.

coded badge see ACCESS CARD.

coded card see ACCESS CARD.

coded signal a method of alerting staff to an emergency on a public address system. In order to avoid panicking members of the public, a pre-arranged apparently routine announcement can be made to advise staff of a fire, bomb threat and so on. Frequently used in large stores, cinemas and theatres.

coherent fibre (bundle) see FIBRE OPTICS.

collapsible post a method of preventing unauthorised vehicle entry or parking by means of a hinged or removable post set in the centre of a parking space or roadway. The post is secured in the upright position by an integral lock or a PADLOCK.

collar 1. the part of a LEVER LOCK KEY which prevents the BIT and SHANK from being pushed too far into the KEYWAY. (See diagram 6 on p. 61) **2.** an ANTI-SCALING BARRIER used to protect trunking or drainpipes.

collector a piece of equipment used in transmitting alarm signals from the premises protected to the CENTRAL STATION. Where LEASED LINE costs are likely to be prohibitive due to the distance from the central station, alarm companies often install a collector (or satellite) at which are terminated a number of alarm lines. The signals from these lines are then sent by a MULTIPLEX SIGNALLING SYSTEM over a single leased line to the central station. The cost of the leased line is then shared among those using the collector.

combinate to set a COMBINATION in a lock or key.

combination 1. the arrangement of CUTS on a key or the TUMBLERS and PINS in a lock
2. a group of numbers representing the BITTING or cuts on a key or the tumblers of a lock
3. a set of numbers or letters to which a KEYLESS COMBINATION LOCK has to be set for it to be opened.

combination change an alteration in the arrangement of the internal parts of a lock or CUTS on a key so as to render the previous key or COMBINATION inoperable.

combination lock a popular term for a KEYLESS COMBINATION LOCK.

commercial security see INDUSTRIAL SECURITY.

computer security the relatively recent practice of applying security techniques to all aspects of computer technology. It is normal to consider computer security as two separate problems. **1.** The physical security of the computer and its associated fittings
2. The security of the information within the system and access to the system from outside terminals.

concealed fixing a feature of high security DOOR FURNITURE where access to fixing screws or bolts is impeded by the design of the item.

concealed shackle padlock a PADLOCK where the SHACKLE portion is entirely hidden by the body of the lock when closed.

concertina wire a modified form of BARBED WIRE, or one of the more modern replacements for barbed wire. The wire is produced in a self-supporting coil and is used for ground entanglements or for ENHANCEMENT of fence or wall tops. Concertina wire is very similar to Dannert wire.

connecting bar a (normally) flat piece of metal fixed to the rear of the PLUG which transmits the turning motion of the plug to the BOLT mechanism in the lock CASE. Sometimes referred to as the tailpiece. (See diagram 12 on p. 71)

connecting screw a screw which passes through the BACK PLATE of RIM LOCKS and holds the CYLINDER in position.

console a piece of furniture into which is set items of security equipment such as alarm control panels, closed-circuit television controls and monitors.

construction master key (system) a method of arranging a MASTER KEY SYSTEM during the building and fitting-out phase which ensures that no keys which will access the finished building are available to contractors. The construction keys can be rendered inoperable without disassembling the locks.

contact device any device, which when actuated operates (i.e. opens or closes) a set of electrical contacts such as a microswitch, relay or MAGNETIC CONTACT.

C

contact microphone a sensor used for the protection of glass windows or safes in the form of a GLASS BREAK DETECTOR or a SAFE ALARM.

contacts the SENSORS normally used in an INTRUDER DETECTION SYSTEM to protect doors and windows are known as contacts. See also MAGNETIC CONTACT.

contingency plan a prepared scheme designed to assist in the efficient organisation of a company or facility in the event of a major emergency. Such plans should be prepared to counter a number of eventualities including major fires, natural disasters, strikes, civil unrest, kidnapping of senior executives, product contamination and major chemical or biological escape. See also CRISIS MANAGEMENT.

continuous wiring see CLOSED-CIRCUIT alarm.

contract security the provision of security services by a SECURITY COMPANY as an alternative to recruiting an IN-HOUSE guard force. Contract security also includes CASH-IN-TRANSIT.

control cabinet see CONTROL PANEL.

control centre see CENTRAL STATION.

control equipment the contents of a CONTROL PANEL or any device intended to regulate the operation of a SENSOR, SOUNDER or SIGNALLING SYSTEM.

control key the key used to install or remove the CORES in a REMOVABLE CORE LOCK.

control key switch a KEY SWITCH mounted on or near the CONTROL PANEL and used to switch the alarm system on and off. It may be equipped with a test position which will indicate whether any sensors are in an ACTIVE MODE (for example, windows left open), and this prevents the system being triggered when it is switched on. It should only be possible to remove the key in the 'on' or 'off' position, not in the 'test' position.

controller see CONTROL PANEL.

control panel in an INTRUDER ALARM SYSTEM, the control panel contains the essential circuitry, relays, power supply (both AC mains and battery back-up) as well as a control mechanism, often in the form of a KEY

SWITCH. The control panel will usually display some form of indication of the status of such functions as power supply, which zones are active and which are isolated, and will probably also have a facility for testing any audible device such as a bell. It is essential that the control panel is itself protected from tampering and is inside the PROTECTED AREA. The control panel is sometimes also known as a controller, control unit or control cabinet.

control unit see CONTROL PANEL.

C

core a complete PIN TUMBLER LOCK MECHANISM (often in the form of a figure '8'). Cores are used in REMOVABLE CORE LOCKS where they can be simply and quickly installed or replaced using a special CONTROL KEY (which is different from the operating key).

Cor Key a proprietary lock system based on the use of magnetic elements whose unique feature is the ease with which individual locks can be reconfigured.

corrugated key a key made of sheet metal where corrugations are pressed longitudinally into the BLADE.

courier an individual who carries valuables or important documents. In the US, sometimes spelt 'currier'.

courier dispatch box see CASH BOX (2).

courier (dispatch) case a strong case for carrying important documents or securities, often made of metal and frequently fitted with a chain which can be secured to the wrist of the COURIER.

CPL Certified Professional Locksmith, the second of three grades of registration awarded by the ALOA.

CPO see CRIME PREVENTION OFFICER.

CPP Certified Protection Professional: a qualification awarded only after a rigorous examination and detailed career review by the Professional Certification Board of the American Society for Industrial Security. Although sponsored by ASIS, membership of that society is not a prerequisite for sitting the examination or obtaining the qualification.

cradle bars a method of securing a LOUVRE WINDOW or FANLIGHT using metal bars in the shape of a flat bottomed 'U'.

crash bar a type of PANIC BAR used to permit swift egress in the event of fire. Certain doors, fitted with crash bars may also be equipped with a SENSOR or LOCAL ALARM (1).

crime prevention officer a police officer (or other individual) who is specially trained in the techniques of the prevention of crime and in particular the communication of such information to the public.

crisis management a concept increasingly popular with major corpor-

25

ations who, as part of their CONTINGENCY PLANS have set up crisis management teams to deal with such emergencies as the kidnapping of a senior executive or the threatened contamination of a product. Such teams must reflect a cross section of the skills relevant to the particular incident and should have the necessary authority to act as required. One US corporation insists that no executives over fifty may be members of crisis management teams because of the inherent stress such positions may cause.

C

Cropwood Conference a conference held at Cambridge University, England in 1971 by the Institute of Criminology. At this conference were laid down many of the guidelines which have since governed the relationships between the private sector security industry and the UK government and police.

cross alarm an ALARM SIGNAL triggered by crossing or short-circuiting any electrical circuit.

cross key see CRUCIFORM KEY.

cruciform key a key whose cross-section is shaped like a cross. This type of key is used to operate multi-pin tumbler mechanisms whose pins are arranged in three or four positions corresponding to the BLADES of the key. Also known as a cross key.

cryptography the science of coding or decoding data so as to prevent it from being read and used by unauthorised individuals. Cryptography is now extensively used in electronic data transmission, particularly from computer to computer in the financial sector. Some high security alarm systems use an elementary form of cryptography in the form of a HANDSHAKE to validate the status of the system and confirm the origin of any actual signal.

CSEPA Central Station Electrical Protection Association: a US-based association representing many CENTRAL STATION-operating companies.

currier see COURIER.

currier case see COURIER CASE.

cut(s) the portion remaining of a key BIT or BLADE after it has been shaped to operate a lock. (see diagram 5 on p. 61)

cyclone fence a term used in some parts of the US to describe a wire mesh fence similar to a CHAIN LINK FENCE but with larger mesh.

cylinder the part of a PIN TUMBLER which contains the PIN TUMBLER MECHANISM which is part of the lock or is attached or connected to the lock at the time of fixing. (see diagram 10 on p. 70)

cylinder collar the metal ring, sometimes decorative, fitted beneath the actual CYLINDER, which protects and conceals the hole drilled to install the lock. Bevelled (and rotating) collars prevent the forcible turning of the whole cylinder SHELL by a pipe wrench. The cylinder collar is also sometimes known as a cylinder rose or cylinder ring.

cylinder guard a hardened metal plate fixed in front of a cylinder lock to prevent wrenching or drilling the CYLINDER. Also known as a (drill-resistant) shield or armoured front. (See diagram 10 on p. 70)

cylinder key a generic term for keys used in disc or PIN TUMBLER or SIDEBAR locks, commonly PARACENTRIC. (See diagram 5 on p. 61)

cylinder lock a complete key-operated unit consisting of a PLUG, SHELL, TUMBLERS, springs and a CAM or CONNECTING BAR or tailpiece. (See diagram 10 on p. 70)

cylinder plug see PLUG.

cylinder ring see CYLINDER COLLAR.

cylinder rose see CYLINDER COLLAR.

cylinder set screw a small screw found in those MORTICE LOCKS and MORTICE SASHLOCKS fitted with a PIN TUMBLER MECHANISM, which prevents the cylinder from turning after installation.

cylinder shell see SHELL.

D

D's, three a maxim used in crime prevention to encapsulate the essential factors – Deter, Delay and Deny. Deter the intruder; if that fails, delay him; should that fail, deny him any advantage.

damped sensor a low sensitivity VIBRATION SENSOR which is normally used at locations subject to substantial WIND LOADINGS. For example on a fence such a sensor would be used in place of a conventional INERTIA SENSOR.

Dannert wire see BARBED WIRE; CONCERTINA WIRE.

data logger see LOG.

day/night switch see AUTHORISED ACCESS CONTROL SWITCH.

dead area an area within a protected location which is not covered by any sensor or piece of surveillance equipment. This may be caused by shadowing from other objects or by incorrect adjustment of sensors or cameras. See also BLIND AREA.

deadbolt the BOLT of a lock which is moved in both the locking and opening directions by the key. The important feature of a deadbolt is that it cannot move until a key is turned and will not respond to a handle. (See diagrams 7 and 12 on p. 67 and p. 71)

deadbolt hole the hole in a STRIKE PLATE into which the DEADBOLT fits when in the SHOT or locked position. In the case of high security locks, the deadbolt hole is often fitted with a BOX as added protection. See also BOX STRIKE. (See diagram 9 on p. 69)

dead ground a military term used to designate a piece of ground which cannot be seen (usually because of contours) from the viewer's position.

deadlatch 1. a NIGHTLATCH which can be DEADLOCKED by turning the key an extra turn
2. a nightlatch which has an anti-thrust slide or bolt to automatically deadlock the LATCH. Often, the LATCH BOLT on such automatic deadlatches is thrown further into the STRIKING PLATE to provide additional security. (See diagram 11 on p. 71)

deadlock a lock, usually fitted with a rectangular BOLT which can be operated only by a key. When the key is turned, the bolt is effectively

28

'deadlocked' and will not move unless the key is turned in the other direction.

debug to search for (or SWEEP) and remove covert surveillance devices (or BUGS).

DED Data Encryption Device.

dedicated line a signal transmission line (which may be a telephone company line) used for no purpose other than connecting an alarm installation to a CENTRAL STATION or other monitoring equipment.

defence in depth the concept of protecting a location by a series of 'layers' of security, using both physical means (walls, fences, locks) and detection equipment. (See diagram 2 below)

defensible space a theory, based on a 1972 book of the same name by US architect and planner, Oscar Newman. Newman suggested that much vandalism and street crime could be eliminated by more attention to the way in which buildings and the space around them were planned. He believed that if the occupants of buildings could somehow gain control over the areas around their homes, this in itself would help to deter crime. His theories, once revolutionary in that they criticised tower blocks and other high-density housing, are now wholly accepted by most authorities.

defensive driving the skill of being able to drive an automobile or other vehicle in such a manner as to avoid putting the vehicle in a position where an accident can take place. A modified form of this skill is taught to those who are required to drive possible terrorist or kidnap targets.

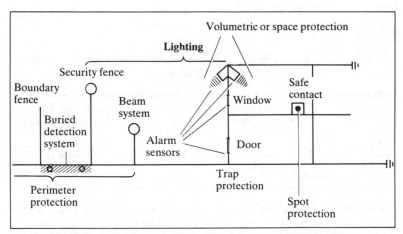

Diagram 2 *Defence in depth: types of detection and protection equipment*

Following such training, it is possible to avoid many of the traps likely to be set by would-be kidnappers. See also EXECUTIVE PROTECTION.

degausser a device used to erase magnetically-coded information.

degraded mode an ACCESS CONTROL SYSTEM which normally works ON-LINE, but has a limited or reduced capability as an OFF-LINE unit is said to be able to operate in a degraded mode. This type of component could also be said to be working as a STAND-ALONE or OFF-LINE CARD READER.

delay circuit a component in an alarm system which allows the person switching the system on or off time to enter or leave the building. Delay circuits are normally used in conjunction with an ALARM BYPASS or SHUNT and are also associated with BELL DELAYS.

deliberately operated device see ALARM STATION.

demand camera see BANK CAMERA.

deposit safe a self-contained safe fitted with facilities resembling those of a bank's night deposit safe. Such safes are fitted with a one-way trap or hopper to permit items to be put into the safe without the need for staff to have access to a key to the door of the safe. Their primary use is to secure valuables outside office hours, and they are much used by hospitals for security of the property of patients who are admitted outside normal working hours.

detachable bit key a key made in two parts, commonly used in high security safes and vault door locks where the key SHANK is very long and thus impractical to carry. To maintain security, the BIT can be removed from the shank and carried on a key ring while the shank is kept near the safe or vault door.

detainer a lock TUMBLER which must be raised or turned to a precise position before the BOLT of a lock can move.

detection circuit a wiring circuit connecting one or more SENSORS or detectors to the control panel of an alarm system.

detection device a SENSOR or other device designed and installed to trigger an ALARM CONDITION in response to an unauthorised entry or attempted entry, or as a result of TAMPERING or the deliberate actuation of an ALARM STATION.

detection field the area of protection provided by a SENSOR in VOLU-METRIC or SPACE PROTECTION.

detection loop see MAGNETIC BURIED LINE SENSOR.

detection range the greatest distance or limit at which a sensor can be activated by an intruder. For example, in the case of PASSIVE INFRA-RED SENSORS, the detection range is the scale of temperatures to which the detection circuit responds.

detector a device which senses a change in state from its pre-set or prearranged reference. When the change of state exceeds pre-set limits, the detector will trigger an alarm. Often used as a synonym for SENSOR. (See diagram 2 on p. 29)

detector contact the component within a SENSOR which is triggered by a change in state and signals this fact to the CONTROL PANEL.

detector lock a LEVER LOCK MECHANISM fitted with a detector lever, constructed so that if any lever is over-lifted by an incorrect key or lock PICK, the detector lever jams in the locked position until it is reset. Resetting is done when the correct key is turned in the reverse direction, after which the lock may be opened normally. Detector locks used often to be specified for securing cabinets or cupboards in which important or classified documents were stored so that any unauthorised attempt to open the cupboard would be noted.

deterrent in industrial security, the term is normally applied to equipment or facilities which will discourage potential or actual criminals from a particular course of action. Thus the presence of an external alarm bell may deter a burglar while a visible closed-circuit television camera in a store may deter a shoplifter.

Detex lock the proprietary name for a type of ALARMED EXIT DEVICE which consists of a compact, surface-mounted box containing a PIN tumbler cylinder, dry battery/cell and a sounder together with an actuating mechanism. A lever protrudes from the box and if this is pushed it withdraws a LATCH BOLT and at the same time activates the sounder. The cylinder lock (which can be fitted to both sides of the door in some units) enables authorised persons to use the door without activating the sounder.

detonator a device for initiating an explosive charge or IED or actuating a mechanical function.

DHI Door and Hardware Institute (US).

dialler see DIGITAL COMMUNICATOR.

differential pressure system see PRESSURE DIFFERENTIAL SYSTEM.

differs the number of different COMBINATIONS available in a lock.

digital communicator a device that automatically dials a pre-programmed telephone number and transmits a digitally encoded alarm message to a CENTRAL STATION or other monitoring unit. Sometimes known as diallers or digital diallers, these units are most often found in smaller retail alarm systems or in larger domestic premises. Normally, a DEDICATED LINE is used and if properly installed the device offers good security. Also known as an electronic digital communicator.

digital dialler see DIGITAL COMMUNICATOR.

digital keypad a KEYPAD with numbered pushbuttons often found in ACCESS CONTROL SYSTEMS.

digital lock a mechanical or (sometimes) electrical lock which operates when a previously encoded sequence is entered via pushbuttons. This type of lock is sometimes known as a Simplex lock after the principal manufacturer.

direct line signalling (equipment) the use of a DEDICATED LINE or lines and the necessary circuitry to connect an ALARM CONTROL PANEL with a CENTRAL STATION or other remote location.

directory fraud a fraud which most commonly occurs when a company receives a document purporting to be an invoice for an unwanted or unauthorised entry in a trade or telex directory. If invoice certification procedures are inadequate, such documents may be passed for payment. To avoid prosecution, some fraudsters have been known actually to publish a single copy of a directory. A similar fraud used to be perpetrated against small businesses when a 'salesman' would canvass for advertising for a non-existent directory and request advance payment.

disaster plan a CONTINGENCY PLAN prepared by an organisation to enable it to deal with major incidents such as fire, flood, hurricane or civil disturbance.

disc see DISC TUMBLER.

DISCO Defence Industry Security Clearance Office: US Department of Defence section which provides security clearances for employees of defence industry contractors.

disc tumbler a type of DETAINER manufactured in the form of a thin metal disc.

disc tumbler lock 1. a thin, flat metal TUMBLER which must be drawn into the cylinder PLUG by the key so that none of its extremities extend into the SHELL, thus permitting the plug to turn
2. a flat DETAINER with a GATE which must be aligned with a SIDEBAR by the proper key.

dishonesty testing a term used to describe various methods of verifying that staff (particularly in the retail sector) are following correct procedures when accepting cash, writing sales slips and ringing. up receipts. See also TEST PURCHASE.

displacement of crime a theory which holds that where the quantity of crime is constant, any successful attempts to improve security and deter crime in one location will simply force the crime to another area.

disrupter a small shotgun-like device used to fire a charge of water, beads or shot at a suspected bomb or IED. The disrupter is intended to fragment the IED before it explodes, enabling the components to be recovered for examination and possible evidence.

DOD US Department of Defence.

dog bolt see HINGE BOLT. (See diagram 4 on p. 51)

door chain a simple piece of inexpensive security equipment which should be fitted to the principal door of all domestic premises. It consists of a short length of strong chain firmly anchored to the door frame; the free end is fitted with a keeper which fits into a metal slot secured to the door. When the chain is engaged, the door can be held partly open, making it possible to speak to a caller who is unable to force an entry. Adequate fixing is essential if the chain is to prove of any value.

door closer a mechanical or hydraulic device for automatically or remotely closing a door or gate.

door furniture the metal items fitted to doors are referred to as door furniture or door hardware. This term includes locks, handles, finger plates, hinges, chains, DOOR VIEWERS, DOOR CLOSERS, stops, and lock or latch striking devices.

door guard see DOOR LIMITER.

door holder a device which holds a door in the open position and is usually electromagnetic. Such devices are often fitted to fire or smoke stop doors and interconnected to the fire detection system. If the system is triggered, the doors are automatically released allowing them to close.

door limiter a simple metal security device which can be fitted to wooden doors in place of a DOOR CHAIN. The limiter, being rigid, offers better security than a chain. Also known as a door guard.

door link a flexible conductor, with a terminal block at either end used to connect a LOCK SWITCH or SENSOR fitted to a door to the alarm monitoring circuitry. Also known as a door loop.

door loop see DOOR LINK.

door phone an electronic communications device frequently found in residential blocks which enables residents to talk to a caller at an outer door and to open such doors remotely.

door strike a plate fitted to a door JAMB or frame intended to receive the LATCH BOLT. Also known as a STRIKING PLATE, the strike will normally also be fitted to receive the DEADBOLT. (See diagram 7 on p. 67) See also ELECTRIC DOOR STRIKE.

door switch a type of mechanical SENSOR fitted to doors which triggers an alarm when the door is opened. Often used in small shops to indicate the presence of a customer.

door viewer a simple, cheap optical device using a wide-angle lens which permits a house occupier to view callers before opening the door. Now also frequently found in hotel guest room doors.

door viewer system a type of closed-circuit television system enabling occupants of premises to view as well as question callers and then remotely to open external doors or gates.

Doppler shift a SENSOR sends out a pattern of energy and when that energy bounces back, the sensor will register the amount of energy returned. If the two amounts vary, it is because something is moving within the energy field. The variation in reflected energy is known as the Doppler shift or effect. This phenomenon applies to all types of ACTIVE SENSOR.

double door holder a mechanical device consisting of a bolt-type mechanism which automatically locks the inactive or passive LEAF of a pair of double doors when the active leaf is closed. This for example permits one fire exit alarm lock (such as a DETEX LOCK) to control a double door.

double drop (system) a method of signalling an ALARM CONDITION, now largely obsolescent, where an incoming signal is 'opened' to create a BREAK ALARM (1) then short-circuited to produce a CROSS ALARM.

double-pole detection circuit a method of HARD-WIRING an alarm circuit using two conductors each in a different electrical state. An ALARM CONDITION or FAULT signal is triggered if the two conductors come into contact with each other.

double supervised (system) a method of alarm supervision used mainly for fire alarms where the power source used for FAULT or TROUBLE SIGNALS is itself monitored against loss.

double throw lock a lock in which the DEADBOLT can be SHOT or extended further into the STRIKING PLATE by a second or additional turn of the key.

dragon's teeth a type of temporary road-blocking device used by the police and military which consists of an expanding metal framework which is fitted with a large number of detachable spikes. When the device is placed on the roadway, any vehicle failing to stop (and passing over the framework) will have enough of its tyres deflated by the spikes to stop it. See also CALTROP, CHEVAL DE FRISE.

drill-resistant pin see PIN TUMBLER MECHANISM.

drill-resistant shield see CYLINDER GUARD.

driver pins the top or upper pins in the cylinder of a PIN TUMBLER MECHANISM. (See diagram 10 on p. 70)

drivers see DRIVER PINS.

dual screw a SUPPLEMENTARY LOCKING DEVICE used to secure wooden casement or sash windows. This simple device utilises a threaded steel locking pin which is inserted into a brass bush or sleeve (already fixed to the window frame). When the steel pin is screwed home using a key-like device, both parts of the window are secured.

dummy cylinder a non-functional replica of a RIM or MORTICE CYLINDER LOCK used for appearances only. Such an item would be used to conceal an old cylinder hole in a door or match another cylinder on say, double doors.

duplex a data transmission system which is able to send data in both directions simultaneously down the same communications link. Also known as full duplex. See also MULTIPLEX SIGNALLING, SIMPLEX.

dwell time the length of time a SEQUENTIAL SWITCHER allows the display of a selected closed-circuit television image on a MONITOR.

E

EAS electronic article surveillance: a term used to describe the whole range of TAGGING systems used to prevent merchandise being removed from sales floors. EAS is also beginning to be used in areas other than retail, in particular in university libraries. The development of paper-thin TAGS will almost inevitably result in a substantial price decrease for this type of equipment.

eavesdropping device any of a range of equipment which can be used to listen covertly to discussions or conversations. Such equipment includes BUGS (1), TELEPHONE TAPS and PARABOLIC MICROPHONES.

echo pulse fault locator see TIME DOMAIN REFLECTOMETER.

EDR Explosive and Drill-Resistant: a term used to describe safes that are designed to resist these forms of attack.

E-Field System a proprietary PERIMETER DETECTION system used to ENHANCE walls and fences with ELECTROSTATIC FIELD SENSORS. The wall or fence is fitted with outriggers along which run two sets of very fine wires: the field wires which emit the field and the sensing wires which receive the return signal. The presence of an intruder within the field alters the pre-set values within the control unit (which are continuously monitored), and any changes beyond a predetermined range will trigger an alarm.

electret cable a weatherproof PVC-covered cable (of similar construction to COAXIAL CABLE) which has properties similar to that of a microphone. When the cable is run along a CHAIN LINK FENCE using plastic ties, it uses this property to pick up any attempt to cut, scale or otherwise evade the fence. When the cable is connected to an appropriate control system it is possible to pre-set certain values which will cause the control unit to ignore activity such as wind noise. Direct monitoring of the signals produced by the cable is possible using headphones or a loudspeaker, and enables a skilled operator to discriminate between different signals produced by various stimuli.

electric door strike a mechanism for locking and unlocking doors, consisting of a mechanical LATCH and an electrically operated solenoid. When the solenoid is energised the latch pin is withdrawn, causing the strike to be released. This leaves the door free to open. See also DOOR STRIKE.

electric field sensor a sensor used in PERIMETER PROTECTION which will detect any change in a predetermined level of an electromagnetic field and trigger an alarm.

electric lock a lock incorporating an electrically operated solenoid or motor to operate the BOLT. Some electric locks also have a manually operated push-button release.

electric vibration device see INERTIA SWITCH.

electromagnetic lock see MAGNETIC LOCK.

electromechanical detection device a sensor which converts mechanical activity into an electrical impulse. An example would be a tilt switch, where movement of mercury from one end of a glass tube to the other makes a connection, allowing current to flow.

electron detection a method of detecting the presence of explosive vapours, based on the principle that molecules of explosive vapour have a high affinity for some electrons and will try to combine with them. In this type of EXPLOSIVE DETECTOR, an inert gas (such as argon) is used to sweep the vapour being tested into a chamber. Any change in the population of thermal electrons is noted and compared with pre-set limits, initiating an alarm as necessary. This type of device is more likely to be accurate than the vapour detector, but it is dependent on an external supply of inert gas and takes longer to warm up as well as being more expensive.

electronic digital communicator see DIGITAL COMMUNICATOR.

electronic guard patrol recording system a modern version of the WATCHMAN'S CLOCK where patrol or tour information is recorded electronically inside a small, pager-like device.

electronic vibration detector a SENSOR using a CONTACT MICROPHONE used for SPOT PROTECTION.

electrostatic field sensor a PASSIVE SENSOR normally used for PERIMETER PROTECTION, particularly on top of walls. Electrostatic field sensors respond to changes in the ambient electrical field of the area covered. See E-FIELD SYSTEM, G-LINE.

ELF European Locksmiths' Federation: a federation of the various national locksmithing associations in Europe.

ELR end-of-line resistor: electronic component used to terminate a SUPERVISED CIRCUIT. The device is used to ensure continuity and to provide a reference point against which changes in values can be measured. Also known as an EOLR.

embossed card a plastic data card, produced with raised characters used in ACCESS CONTROL SYSTEMS or for imprinting documents. In the case of access control, such cards are called HOLLERITH CARDS, (although strictly speaking, Hollerith cards have holes punched in them). It is possible to emboss many ACCESS CARDS so they can be used like credit cards for such functions as tool issue, staff purchases or payroll. In the British Army, the standard IDENTITY CARD has embossed details such as name, rank and service number, which enables the card to be used to prepare passenger manifests.

embossing machine a machine used for the production of EMBOSSED CARDS by means of pressure and heat. All credit cards and CHEQUE CARDS are produced in this way.

emergency master key a key which will operate all the guest room locks in a hotel MASTER KEY SYSTEM, even when such locks are secured from the inside. This key will often act as a SHUT-OUT KEY and can be used to render the lock inoperative to all other keys.

emergency ventilation system a mechanical device using either a fan or mechanical air mover and metal tubes to provide non-toxic air to a VAULT or REFUGE. Speech is also possible via the tubes and with some devices, a limited amount of food or water can also be transferred through the tubes in the event that someone is trapped in a vault for some time.

emotional stress monitor see POLYGRAPH.

encode see COMBINATE.

encoded card see ACCESS CARD.

encoder a device used to input information into an ACCESS CARD, bank or credit cards.

endoscope a small diameter FIBRE OPTIC instrument used for visually examining the interior of objects such as the body panels of vehicles when searching for contraband or explosive devices.

engineer reset a feature often found in INTRUDER ALARM SYSTEMS fitted with direct connections to central stations. In the event of an alarm, the system can only be reset by a service engineer. This feature is mandated by many standards and alarm specifying bodies.

enhance to upgrade or improve the physical security of a location. An example of enhancement would be the addition of BARBED TAPE to a perimeter fence. The barbed tape is referred to as enhancement. TARGET HARDENING normally involves a great deal of enhancement.

enrolment the action of entering an individual's details into an ACCESS CONTROL SYSTEM's memory either by encoding an ACCESS CARD or by directly keying data into a computer.

entry code an alphanumeric sequence that has to be punched into a KEYPAD (to operate an ACCESS CONTROL SYSTEM) as a secondary check on the cardholder's identity. Entry codes are used in conjunction with ACCESS CARDS.

entry control system any combination of hardware and circuitry which is used to regulate entry or egress in any location. Often used as a synonym for ACCESS CONTROL SYSTEM.

entry delay a predetermined period during which an alarm system will not be triggered to permit authorised personnel to enter a location and isolate the system. See also ALARM BYPASS, DELAY CIRCUIT, SHUNT.

entry route the normal route between the exterior and CONTROL PANEL of a PROTECTED AREA. This route must be followed by the KEYHOLDER when switching off the system.

entry system This term is often used as a synonym for DOOR PHONE system. See also ENTRY CONTROL SYSTEM.

EOD Explosive Ordnance Disposal: the formal military term for bomb disposal. The term is used to encompass the safe disposal of military explosives as well as terrorist devices. See also IED.

EOLR see ELR: END-OF-LINE RESISTOR.

eosine a moisture-activated, VISIBLE STAINING COMPOUND producing a distinctive red colour on skin contact.

EPIC Ex-Police in Industry and Commerce: a UK association of former police officers working in industrial security.

escutcheon (plate) a small, swivelling plate fitted over the keyhole of a lock, which is often used to prevent the entry of dirt or moisture into the lock case. Often ornamental or decorated.

espagnolette a device often used for fastening double-leaf windows, particularly french windows, consisting of a single, full-height bolt operated by a central handle, used to secure the top and bottom of the window.

espionage the (illegal) gathering of secret information, normally by one state against another. Such activities, when carried out in the commercial sector, are known as INDUSTRIAL ESPIONAGE.

evacuation signal any emergency signal, whether a CODED SIGNAL, speech or via a SOUNDER, which is used to advise personnel that it is necessary to leave a building. In the case of a fire, in many buildings, sounding a GENERAL ALARM (that is by ringing all the fire bells or other devices) is used to signal an evacuation. In other areas, pulsed or intermittent bells are used. In locations such as hospitals where the occupants may panic, a coded voice message will often be employed to give an initial warning to staff.

EVD Electronic Vibration Detector.

event printer see LOG.

executive protection a term used as the non-governmental equivalent of close protection. Both terms encompass the various activities involved in preventing any harm to senior management, dignitaries or politicians. Executive protection may include the actual bodyguarding of the person being protected, driving him and providing physical protection for him and his home, family and work place. Among the areas to be covered when planning executive protection are evasive and DEFENSIVE DRIVING techniques for his chauffeur; the physical security of his home and office; construction of a REFUGE and a detailed analysis of his lifestyle. Lifestyle analysis will show up possible weaknesses in the security plan which would be spotted by kidnappers. The need for KIDNAP AND RANSOM insurance may also form part of the overall strategy. A thorough plan will also anticipate possible failure and teach the executive survival techniques to enable him to live through any hostage-taking or kidnapping. Executive protection is a highly specialised field and one most security practitioners should leave to those trained and experienced in it. See also CRISIS MANAGEMENT.

exit alarm an alarm fitted to a fire exit door which is designed to trigger an alarm when the door is opened.

exit button a device used to actuate an electrically operated door to permit egress.

exit delay a predetermined period during which an ALARM SYSTEM will not be triggered. Used to permit authorised personnel to switch on an alarm system and leave the PROTECTED AREA.

exit device see CRASH BAR, PANIC BAR.

exit route the normal route from a CONTROL PANEL to the exterior of a PROTECTED AREA. This route must be followed by the KEYHOLDER after switching on an alarm system.

explosion-suppression blanket see BOMB BLANKET.

explosive detector an electronic device which will indicate the presence of certain explosive compounds by analysis of the vapour emitted or by ELECTRON DETECTION. Such devices may be hand-held (in the case of the vapour detector), or semi-portable (in the case of the electron detector) for searching personnel and baggage, or may be incorporated in a conveyor system for high-volume screening of freight.

external audible alarm a SOUNDER together with a power supply and associated equipment mounted on the exterior of a PROTECTED LOCATION. It is essential that the equipment cabinet or bell box is fitted with an ANTI-TAMPER DEVICE.

eye cable a steel cable finished with a loop or loops used to secure gates or other equipment by means of a padlock.

Eye-Dentify a proprietary identification system based on retinal scanning techniques, primarily used for ACCESS CONTROL. Retinal scanning is one of a number of BIOMETRIC SENSOR systems.

F

f number see APERTURE.

face plate see FOREND.

facility code a code associated with an ACCESS CONTROL SYSTEM or ALARM SYSTEM, which identifies a user or a specific location.

factory police an obsolete term for an industrial security force or its personnel.

fail-open see FAIL-SAFE.

fail-safe a device or system that initiates an alarm in the event of a power failure or equipment malfunction is said to fail-safe. In some buildings, fire doors are held locked electrically, and in the event of a fire alarm being triggered, or if there is a power failure, the doors are released. These are also said to 'fail-open'.

fail-secure a device that in the event of a power failure automatically secures itself or remains in a locked or closed position is said to fail-secure.

false alarm any spurious or unwanted alarm, especially one caused by a defect in a sensor or the system. See also NUISANCE ALARM.

fan an ANTI-SCALING BARRIER used to prevent access to balconies or walkways. Normally made from spiked rails welded together in a fan-shape.

fanlight a window set above a door, which is normally semi-circular in shape and often fixed. See also TRANSOM WINDOW.

FAR False Alarm Rate: the frequency with which an ALARM SYSTEM actuates as a result of spurious or unwanted alarms.

fault a term used to describe an abnormal condition in any fire or security system. Most CONTROL PANELS actually include fault indications to distinguish system failures or errors from actual alarms. Faults must be cleared before the system can be restored to normal operation. Most intruder systems will not allow themselves to be switched on while there is a fault present. In SUPERVISED CIRCUIT systems, the presence of an irregularity in any part of the system will result in a fault's being triggered.

FCRA Fair Credit Reporting Act: US legislation which (among other things) mandates the way in which credit checks can be performed. Has an implication in respect of BACKGROUND SCREENING.

fence 1. a flexible barrier, normally a CHAIN LINK FENCE or similar material fixed to posts, used to protect a site or mark a boundary. Boundary fences act simply as property markers while security fences are invariably ENHANCED with some form of barbed material and often fitted with a FENCE SENSOR system (See diagram 2 on p. 29)
2. a slang term for a person who habitually deals in stolen property.

fence sensor the part of a PERIMETER PROTECTION SYSTEM which is actuated if an intruder touches or comes close to a protected fence. See also ELECTRIC FIELD SENSOR, INERTIA SENSOR, TAUT WIRE SYSTEM, VIBRATION SENSOR.

FERIT Far East Regional Investigation Team: a part of the International Maritime Bureau based in South East Asia with a special responsibility to investigate MARINE FRAUD.

ferrous metal perimeter alarm (system) see ELECTRIC FIELD SENSOR.

fibre optics a method of light transmission using specially manufactured glass fibres. A group of fibres, known as a bundle, will consist of many thousands of individual strands. It is possible to use such bundles for the transmission of images (using coherent fibre bundles), or for signal transmissions in the form of modulated light (using non-coherent fibre bundles). A high degree of security is obtained from this type of equipment and there are likely to be substantial developments and expansion in the use of fibre optics in the security field.

fibre optic tape a method of PERIMETER defence in which a non-coherent fibre optic bundle is bonded to a length of metal tape. The tape (which may be barbed or plain) is then used to ENHANCE a fence or wall. Any distortion or breaking of the tape (as would take place in an intrusion) results in an attenuation of the light passing through the fibre bundle. This is sensed by the CONTROL EQUIPMENT which triggers an alarm. Very low false alarm rates have been reported for this type of system.

fidelity bonding see BONDING.

field of view see FOV.

field wire in an ELECTROSTATIC FIELD SENSOR, there are two sets of wires. The field wire actually emits the signal while the sensing wire receives the reflected signal. See also E-FIELD SYSTEM.

FIISec Fellow of the Institute of International Security see IISEC.

final exit a door to premises which is used for leaving the location after an ALARM SYSTEM has been switched on.

final exit lock a lock used to secure a FINAL EXIT door, usually fitted with a LOCK SWITCH or controlled by a SHUNT LOCK or SHUNT SWITCH.

Fingermatrix a proprietary ACCESS CONTROL SYSTEM using a low power laser to read fingerprints. See FINGERPRINT READER.

fingerprinting a method of identification using the unique pattern of loops and ridges found on human fingers.

fingerprint reader a high security identification and ACCESS CONTROL SYSTEM which scans the tip of a finger or thumb by means of a special reader and when coupled to the appropriate interface equipment, permits access if the configuration scanned matches previously stored information. See also BIOMETRIC SENSOR.

FIPI Fellow of the Institute of Professional Investigators see IPI.

fireman's key a key which overrides the normal working of a lift or elevator and brings one or more cars to the ground floor, disabling the automatic response to floor call buttons.

fitch catch a simple, non-locking catch used to secure vertically-sliding SASH WINDOWS, also known as a cam catch.

flat key a key made of steel which has no grooves or corrugations, frequently used for cabinets, lockers and luggage.

float glass see ANNEALED GLASS.

floor limit in retail security, the amount of a credit card transaction which does not require approval from a credit control department or authorisation from the credit card company.

floor pad see PRESSURE PAD.

fluorocene an INVISIBLE STAINING COMPOUND used in the detection of theft or in cases involving tampering with property. It is very persistent and will mark the skin, becoming visible only under ultraviolet light.

flush bolt see FLUSH-MOUNTED DOUBLE-DOOR BOLT.

flush-mounted double-door bolt a simple non-locking device fitted on the top and bottom edge of the active or first-closing LEAF of a pair of doors or french windows. Also known as a flush bolt. (See Diagram 3 on p. 45)

fluttering a slang term used to describe the process of subjecting an individual to a POLYGRAPH examination.

FOC Fire Office's Committee: a UK-based insurance-sponsored standards and testing body. Now under the authority of the LOSS PREVENTION COUNCIL. *Note*: often incorrectly written as Fire Officer's Committee.

foil a thin, narrow metallic ribbon normally used for protecting windows in low security applications. The foil strip, often self-adhesive, is fixed to

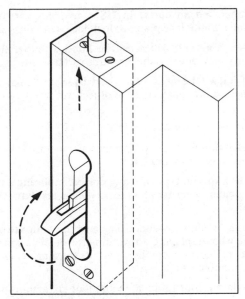

Diagram 3 *Flush bolt (double door)*

F

the inner surface of the glass. If a window is broken, the foil breaks, opening a circuit which triggers the alarm.

foil connector an electrical connector or terminal used to connect the ends of FOIL circuits.

foil crossover an insulated bridge which allows FOIL to pass over window dividers or other conductive materials.

foil take-off block a terminal block used to interconnect the ends of a FOIL run to the alarm circuitry.

follower that part of a lock or LATCH which contains the square SPINDLE HOLE and allows the handle to withdraw the LATCH BOLT. (See diagram 9 on p. 69)

foot candle a measurement of light (now only US). One foot candle is the amount of light which falls on a surface one foot from a light source rated at one international candela/lumen. One foot candle is equal to 10.764 LUX.

foot switch a BANDIT ALARM or PANIC BUTTON actuated by foot. Also known as a kick bar.

forend the part of a lock through which the bolt protrudes and by which the lock is fixed to the door. On security locks there are usually two forends, an inner and an outer. In US usage, this part of the lock is sometimes known as the face plate, or lock face. (See diagram 9 on p. 69)

fourway lock a RIM LOCK which can be fitted to either the left or the right of a door which opens either inwards or outwards.

FOV Field of View: the area supervised by a particular closed-circuit television camera and in particular the object size which can be portrayed on a MONITOR.

FPA Fire Protection Association (UK).

FPS Fence Protection System a proprietary name for a PERIMETER PROTECTION SYSTEM based on ELECTRET CABLE.

frame fastener a special type of fixing device, replacing a nail or screw, which offers greater security when a new, heavier duty, door frame is being installed.

freeze frame the ability to render an image on a MONITOR motionless. Normally done when replaying a recording for detailed examination of a scene or incident. Recent technology permits freezing of real-time images using built-in circuitry.

frequency division multiplex a method of data transmission where multiple signals share a single transmission band by the allocation of specific channels within the band to each separate signal.

furniture see DOOR FURNITURE.

G

GAI Guild of Architectural Ironmongers (UK).

gas detector a portable or fixed electronic device which responds to the presence of a specific gas and initiates a LOCAL or GENERAL ALARM.

gate the slot in a lock LEVER through which a BOLT STUMP passes when the bolt is operated. (See diagram 7 on p. 67)

gatehouse a small building situated at the entrance to a site for the use of security staff. Gatehouses control site entry and exit and are often equipped for other functions such as KEY CONTROL and alarm monitoring. See also GUARDHOUSE, LODGE.

gatelodge see LODGE.

gating a method of reducing false alarms, mainly in fire detection systems. When an alarm is gated, a pre-selected threshold of alarm stimulus causes the remainder of the detection devices in that zone automatically to reduce their sensitivity. After a predetermined period has elapsed, the detectors return to their normal level of sensitivity and if the stimulus which produced the gated alarm is still present, a full alarm will be triggered. This circuitry can help to reduce the number of spurious alarms caused by, for example, pipe smokers.

general alarm an alarm which is sounded throughout a site or building. In the case of a fire alarm, a general alarm is one where all sounders are actuated to indicate the presence of a major fire or other emergency. See also CODED SIGNAL, EVACUATION SIGNAL.

geophone a type of sensor used in a SEISMIC DETECTION SYSTEM which responds to the vibrations produced by footfalls or vehicles. See also AUDIO DETECTION.

georgian glass a sandwich of glass and rectangular wired mesh, normally 6 mm thick. It is intended for use primarily to provide a degree of fire resistance in the VISION PANELS of fire doors and has little security value, although it is SHATTERPROOF. Strictly speaking, it is the rectangular mesh which is 'georgian' – other types of wired glass may have hexagonal or diamond shaped mesh. To be classified as 'fire resistant' by UL, such glass must be at least 0.25 inches (6.4 mm) thick.

georgian wired glass see GEORGIAN GLASS.

ghosting the term used to describe the appearance of multiple images on a closed-circuit television MONITOR resulting from video transmission interference.

ghost worker see STRAWMAN.

glass bolt a simple securing device used as a means of controlling a fire or emergency exit. It consists of a spring-loaded BOLT held in the SHOT position by a hollow glass tube. When the tube is smashed (usually by means of a small hammer which is fixed to the device by a chain) the bolt springs open, allowing the door to be used. One model of this device allows the bolt to be smashed from either side of the door using a thin metal rod wrapped around the glass tube.

glass break detector an electronic SENSOR, normally fitted near windows, which is designed to respond to the sound frequencies produced by breaking glass. A similar result can be obtained by fitting an INERTIA SENSOR directly to the sheet of glass to be protected rather than a glass break detector but while less expensive, the inertia sensor has the disadvantage of being more prone to FALSE ALARMS caused by the wind or accidental knocks to the glass.

glass brick a hollow translucent block (either plain or patterned) used in walls and floors to provide natural illumination. Glass bricks are frequently found set into pavements or sidewalks to provide daylight in basements. Unless carefully specified and skilfully installed, glass bricks offer poor resistance to attack.

G-line a proprietary PERIMETER DETECTION SYSTEM based on the passive ELECTROSTATIC FIELD SENSOR. See also E-FIELD SYSTEM.

GLS a conventional filament lamp. Rarely used in security lighting as it produces relatively low light levels per watt and has a short life.

GPBTO General Purpose Barbed Tape Obstacle: US Army trials indicated that the West German BARBED TAPE did not offer any significant advantages over barbed wire and this led to the development in 1970 of what has become known as GPBTO. The tape is produced in coils and is usually made from stainless steel manufactured with very long (65 mm/ 2.5 inch), sharp spikes or barbs. Type 1 tape consists of two interwoven helical coils and as it is intended as a temporary barrier (primarily for military use) it is not fabricated from stainless steel. Type 2 is identical to Type 1 except that it is fabricated from stainless steel. Type 3 consists of a single coil of stainless steel tape and is intended primarily to ENHANCE fences at high security locations. Barbed tape is also produced in large coils under the proprietary name of 'INSTANT BARRIER' which is used for temporary protection of locations or to provide additional perimeter protection.

Grade AA/A/B Central Stations a classification system adopted by UL for proprietary CENTRAL STATIONS. (See Appendix 2 on p. 134)

Grad. IISec Graduate (member) of the Institute of International Security see IISEC.

grand master (key) see MASTER KEY.

graphic annunciator panel see MIMIC PANEL.

graphology the science of handwriting analysis. It is reported that graphology is being increasingly used in staff selection and a number of large corporations have reported that it provides an important adjunct to BACKGROUND SCREENING.

great-grand master (key) see MASTER KEY.

grid sensor a method of protecting openings such as skylights and cable ducts by an arrangement of wire mounted on a light frame. Attempts to penetrate the opening result in damage to the wire which triggers the alarm. Other methods of protecting this type of location are TUBE AND WIRE and LACING.

guardhouse an alternative term for GATEHOUSE – probably from the military usage although some authorities would suggest that a gatehouse is always at the entrance to a facility, while a guardhouse may be located elsewhere on a site.

guardian key see SAFE DEPOSIT LOCK.

G

guard service the provision of security guards by a CONTRACT SECURITY company.

guard tour see WATCH TOUR.

H

hacker a term originally applied to a computer enthusiast but now used in a derogatory sense to describe an individual who habitually tries to access computer systems either for the challenge involved or for mischief-making.

Hailey bridge an electrical device which processes VOICE-GRADE MULTIPLEX SIGNALS between an ALARM CONTROL unit and a CENTRAL STATION. It is used primarily to maximise the use of circuits.

hand geometry reader a BIOMETRIC SENSOR used in ACCESS CONTROL which is capable of electronically measuring the length of each finger and comparing the digitised result with stored data. See also IDENTIMAT.

hand-held a term used to describe portable (as opposed to fixed or installed) equipment.

hands-free access control (system) an ACCESS CONTROL SYSTEM which does not depend on the use of encoded cards or the entry of data into a KEYPAD. This is a major area of security technology development and uses the PROXIMITY READER, which is able to scan a TOKEN or 'active card' (both of which contain a miniature transmitter) at a distance of up to a metre. This means that an individual wearing or carrying an appropriately coded token can approach a proximity reader and, without having to stop and insert a card or key-in a number, a door or other barrier will be automatically opened. Improvements in microchip technology have meant a decrease in the size of tokens together with a substantial increase in battery life (10 years being common). Active cards normally contain a battery, or else derive their power by means of inductive coupling from the proximity reader. This type of equipment was pioneered some time ago under the proprietary name of Mastiff Systems but it is interesting to note that much of the development in this area is now being carried out by companies who originally entered the security industry in the EAS area.

handshake a term used to describe the encoded electronic verification procedure between a transmitter and receiver. High-security INTRUDER ALARM SYSTEMS are normally provided with this feature to validate any signals transmitted to a CENTRAL STATION. See also CRYPTOGRAPHY.

hardware 1. metal door and window fittings, DOOR FURNITURE

2. the actual circuitry and peripherals of an electronic control system (as opposed to the SOFTWARE).

hardwired a term used to describe a conventional alarm or other system that uses cables or wiring to connect individual sensors and other components to the CONTROL PANEL. The term is used to distinguish this type of arrangement from other systems using radio, MICROWAVE SENSORS, or FIBRE OPTICS in place of wire or cable.

hasp and staple the two parts of a fastening used in conjunction with a PADLOCK to secure a door or gate. The hasp is a metal bar, hinged at one end with a slot through which the staple, a curved metal loop, protrudes. The padlock is then fixed through the staple securing the hasp.

heat detector see THERMAL SENSOR.

heel the end of the shackle of a PADLOCK which is hinged to the case.

high durability card an ACCESS or IDENTITY CARD constructed of more durable material than the PVC normally used. Such materials include polyester, mylar and resin/polyester.

hinge bolt a permanently fitted metal BOLT which protrudes from the hinge side of a door and fits into recesses cut into the frame. This provides greatly improved resistance to forced entry at little cost. Also known as a dog bolt. The US device known as a 'pinned hinge' is similar. (See diagram 4 below)

hold-up alarm a manually operated ALARM STATION located where it can be operated surreptitiously in the event of a robbery.

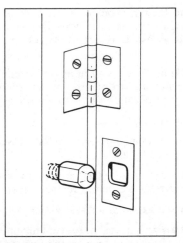

Diagram 4 *Hinge bolt*

51

hold-up camera see BANK CAMERA.

hold-up switch the operating mechanism for a HOLD-UP ALARM. Often in the form of a FOOT SWITCH or similar concealed device connected to a PERMANENT CIRCUIT. Hold-up switches are also used to activate ARMOURED SHUTTERS as well as BANK CAMERAS and closed-circuit television cameras (and recorders).

Hollerith card a type of ACCESS CARD utilising punched holes or embossed markings. Normally found in low security applications, for example, car park barrier control.

hollow post key see BARREL KEY.

Home Office panel an alarm monitoring device which at one time was the only type of equipment which could be installed within police stations in the UK. Thus all alarm systems which were required to be connected to the police had to be capable of being interfaced with this type of equipment. Following changes in police policy, it is no longer possible to connect most commercial or residential intruder alarm systems to police stations. A similar device in the US is known as a police station unit.

The device described above should not be confused with the Home Office Crime Prevention Panel which is an advisory body representing all parties with an interest in the prevention of crime.

hook and eye a simple method of 'securing' sliding doors or cupboards utilising a thin metal hook fastened to a screw eye. The hook portion is inserted into a wire staple. Has no security value.

hook bolt a DEADBOLT, sometimes curved, on locks designed to secure sliding doors and collapsible metal gates.

hook bolt mortice lock a type of MORTICE LOCK where the BOLT is replaced by a flat claw or hook. The hook lock is used to secure sliding doors and in some cases as a SUPPLEMENTARY LOCKING DEVICE for securing windows.

horn 1. a noise-making device used in alarm systems
2. the part of a microwave transmitter which sends and receives the signal. So called as it physically resembles a horn.

horse irons see CALTROP.

house detective an obsolescent term (mainly US) for a plain-clothes security officer employed by an hotel.

huck pin an irregularly serrated-edged top or DRIVER PIN in a PIN TUMBLER LOCK used to increase PICK resistance. See also MUSHROOM, SPOOL PIN.

hydraulic detector see PRESSURE DIFFERENTIATING SYSTEM.

hydraulic vehicle barrier a device to prevent unauthorised entry or exit. There are two principal types of barrier; the first is a simple arm or pole mounted on a pillar. The second type, sometimes known as a 'road blocker' or 'rising step barrier', is normally wedge-shaped and of massive steel construction. The top of the barrier is flush with the ground permitting passage of vehicles when the unit is 'open'. Hydraulic pressure is used to raise the barrier into the 'closed' position where the wedge presents a substantial obstruction to passage – usually between 0.5 m/1.5 feet and 1 m/3 feet high. When correctly designed and installed this type of obstruction can provide positive security against the entry of would-be suicides driving explosive-laden vehicles. See also VEHICLE BARRIER.

I

IAAI Inspectors Approved Alarm Installers (UK).

IAASP International Association of Airport and Seaport Police.

IACCI 1. International Association of Computer Crime Investigators **2.**International Association of Credit Card Investigators.

IACLEA International Association of Campus Law Enforcement Administrators.

IAHS International Association for Hospital Security.

IAPSC International Association for Professional Security Consultants.

IAS INTRUDER ALARM SYSTEM.

IASCS International Association for Shopping Center Security (US).

IATB International Aviation Theft Bureau.

ID Card IDENTITY CARD.

Identimat a proprietary name for an ACCESS CONTROL SYSTEM using a HAND GEOMETRY READER.

identity card a card, usually encapsulated in plastic often containing a photograph used as means of identifying employees or others as having specified rights of entry into a particular location. See also PHOTO IDENTIFICATION, BADGE.

IED improvised explosive device: the military and police term for a bomb or similar device manufactured from commercially available components.

IFE Institution of Fire Engineers (UK).

I. Fire E. see IFE.

IFPO Institute of Fire Prevention Officers (UK).

IHSSF International Healthcare Security and Safety Foundation.

IIRSM International Institute of Risk and Safety Managers.

IISec The Institute of International Security: a professional body dedicated to the improvement of industrial security standards in the UK and elsewhere. Its primary functions are to promote its own examinations

54

and it provides the only UK-based professional security qualification. There are three levels of membership: Graduate, Member and Fellow. The first two grades require the passing of an examination while Fellowships are awarded on presentation of a thesis on a security topic. The Institute is affiliated to the IPSA.

illegal abstraction a term used to describe the theft of electricity and/ or gas, normally by tapping into a supply cable or bypassing a metering device. This crime is also known as theft of service, a term also used to encompass illegally obtaining telephone service.

image intensifier an electronic device originally developed for the military that provides a night vision capability. The equipment works by electronic amplification of available light. These devices are sometimes known as 'Starlightscopes' after one of the first models.

IMB International Maritime Bureau: an international association of shipping companies, shippers, ports and insurers, one of whose primary objects is the prevention and detection of MARINE FRAUD.

impact strength the measure of shock resistance of materials normally measured in Newtons per unit area. Used to specify vehicle barrier strength.

impedance matching the arrangements within an electronic circuit which adjust the impedance of an AC load to the required values of the system.

improvised explosive device see IED.

in alarm see ALARM CONDITION.

incendiary device a (usually) improvised mechanism designed to start a fire. Such devices are frequently used by terrorists, particularly when explosives are difficult to obtain.

incident light a high-powered flood or spotlighting system usually mast-mounted on a vehicle, used by police and emergency services to provide illumination at the scene of an accident or other incident.

individual keying a method of KEYING in which each lock or cylinder has its own unique key.

industrial espionage the acquisition of product, marketing or other commercial intelligence by unorthodox or illegal means. In some quarters, industrial espionage is known euphemistically as 'aggressive market research'.

industrial security the practice of protecting premises, plant and personnel against external and internal threats. A term often used to encompass all aspects of non-governmental security work. Industrial security is sometimes known as commercial security.

inertia sensor a type of sensor useful in a number of security applications. Inertia sensors operate on the principle that matter remains in a state of rest. They normally consist of a metallic ball or other solid shape in a cup. The surface of the ball and the cup are in electrical contact, completing a circuit. If a fence or other object to which the sensor is attached moves, the cup moves too, but the ball tends to remain at rest. This breaks the electrical circuit and triggers an alarm. Inertia sensors or inertia switches as they are sometimes known are relatively cheap but are prone to false alarms unless they form part of a DAMPED SENSOR. Also known as an electric vibration device.

inertia switch see INERTIA SENSOR.

infrared that range of frequencies just below the visible red spectrum. Infrared is used extensively in security technology. See ACTIVE INFRARED, PASSIVE INFRARED.

infrared beam (system) a device using an active infrared BEAM to provide TRAP or PERIMETER PROTECTION.

Infrared beams may be used externally (when they are sometimes known as an infrared fence) or internally. The beams used in anything other than basic domestic protection should be pulse-modulated to prevent an intruder mimicking a beam with a torch or flashlight.

infrared card see OPTICAL DENSITY CARD.

infrared card reader an ACCESS CONTROL component using an INFRARED BEAM to scan OPTICAL DENSITY CARDS.

infrared lamp a light source which produces INFRARED.

infrared lighting the use of INFRARED LAMPS, normally in conjunction with closed-circuit television to provide discreet or covert night-time surveillance of high security locations.

infrared sensor any SENSOR which uses INFRARED. See ACTIVE INFRARED, PASSIVE INFRARED, PULSED INFRARED.

in-house (security) a term used to describe the system whereby a company or organisation recruits its own security force, rather than hiring personnel from a CONTRACT SECURITY or proprietary security company.

in-line phone tap a method of intercepting telephone communications in which a BUG is placed in or on the telephone line rather than inside the telephone instrument or exchange equipment.

inner forend see FOREND.

inside control a term used to describe a locking device which is usually operated from inside a building.

inside perimeter a line of protection within or close to a PROTECTED AREA passing through possible entry points such as doors, windows and skylights and including more especially cable tunnels, culverts, drains and water pipes.

Instant Barrier a proprietary name for large diameter coils of BARBED TAPE which can be speedily deployed from wheeled containers to provide temporary protection. Instant Barrier is also used in controlling civil disturbances.

integrated system the practice (usually only applying in large buildings or very large developments) of using common facilities for the monitoring of fire, security, building automation and plant alarms. Such systems offer advantages of economy of scale and lend themselves to microprocessor control.

intelligent alarm system an INTRUDER ALARM SYSTEM in which there is a two-way flow of information between the SENSORS and the CONTROL PANEL. This facilitates checking of the sensors, as in such systems a self-testing circuit sends out a regular pulse to each sensor. This checking procedure is also used if a sensor transmits an alarm. Before the CONTROL PANEL triggers an alarm, it can interrogate other nearby sensors to see if their output confirms the alarm status. In addition, with intelligent systems it is possible to reduce sensor levels during the day to avoid false alarms.

Interception of Communications Act 1985 a piece of UK legislation which (among other things) makes it a specific criminal act to TAP telephones and other telecommunications links, or to intercept mail.

interlocked doors two or more doors which are connected by an INTER-LOCK SWITCH which makes it impossible to have more than one door open at a time. Often used in accessways to VAULTS or computer rooms and also known as an airlock or mantrap.

interlock switch a switching device which prevents a function being carried out until some other action has taken place. For example, where a room is protected by a gaseous, total-flooding fire extinguishing system, it would be possible to interlock the door-securing mechanism with the fire system control. In this case, the door lock will not function unless the fire system is isolated (and therefore incapable of discharging gas while the room is occupied or the door unlocked).

internal audible alarm a SOUNDER intended to operate within a PROTECTED AREA during the ENTRY and EXIT DELAY periods or to alert staff to the presence of a FAULT or other situation requiring rectification.

internal casement lock a SUPPLEMENTARY LOCKING DEVICE used to secure wooden or metal windows by shooting a BOLT into the frame.

intruder alarm system an electrical or electronic installation consisting of SENSORS, wiring as well as control and monitoring equipment intended to detect and report the presence, entry or attempted entry of unauthorised persons into a PROTECTED AREA. Also known as an intrusion alarm or intruder detection system, and more commonly referred to as a burglar alarm.

invisible staining compounds one of various types of chemical stain which are not visible in normal light, but only under, for example, ultraviolet. Such chemicals are used in theft investigations to determine whether or not suspected individuals have handled marked items. VISIBLE STAINING COMPOUNDS are also available. See also FLUOROCENE.

IOSH Institute of Occupational Safety and Health (UK).

IPI Institute of Professional Investigators: a UK-based professional body intended to support ethical and professional conduct among those working in the investigation field. It conducts exams and awards membership at various grades.

IPSA International Professional Security Association: a UK-based society intended to promote the status and interests of all of those working in the security field. IPSA arranges training courses at all levels including correspondence courses for the IISEC examinations.

iris the component part of a camera which controls the actual diameter of the APERTURE and thus the amount of light entering the camera.

ISCPP International Society of Crime Prevention Practitioners.

ISIT see SIT.

J

jalousie see LOUVRE WINDOW.

jamb a door post or that part of the door frame which is vertical.

jamming 1. the (usually) deliberate blocking of a radio, frequency by electrical signals to prevent proper reception
2. An attempt to bypass a circuit or device by introducing a false or spurious signal.

jemmy 1. (noun) a slang name for a crowbar
2. (verb) to force doors etc open using a crowbar.

jimmy see JEMMY.

joystick an electromechanical device used to input data to a computer or to control the PAN AND TILT motions of a closed-circuit television camera.

judas gate see WICKET GATE.

judas hole a DOOR VIEWER or spyhole.

Judges Rules see PACE.

junction box an electrical accessory consisting of a metal or plastic box in which is fitted a terminal block. Junction boxes are used to facilitate the speedy joining of two or more conductors.

J

K and R see KIDNAP AND RANSOM.

key in addition to the usual type of metal key familiar to all, the term is also used to cover plastic, composition and even paper or foil devices, all of which fulfil the essential role of operating locking or securing devices. (See diagram 6 on p. 61)

key box a steel or wooden box specifically designed for the secure storage of keys. The same term is sometimes used for a small, glass-fronted box in which a single door key is hung. The box is then wall-mounted near a locked door. The supposed rationale is that in the event of a fire, the door can be opened by smashing the glass and obtaining the key. This is a dangerous practice and even contrary to law in some places.

keycard see ACCESS CARD.

key control 1. a method of ensuring that keys are issued or made available only to persons authorised to receive such keys
2. a systematic organisation of keys and key records.

key cuts see BITTING.

keyed-alike systems an arrangement of locks and cylinders which have the same COMBINATION so that a single key will operate all of them.

keyed-different system the opposite to a KEYED-ALIKE system; a different key is needed to operate each lock or group of locks.

keyholder the person in a company or organisation nominated to be contacted by the police or a CENTRAL STATION in the event of an alarm actuation or other incident. The word is also used to describe the individual responsible for SETTING and unsetting an ALARM SYSTEM.

key holding (service) a system where a CENTRAL STATION or CONTRACT SECURITY company is nominated as the KEYHOLDER for a company and arranges to dispatch a security officer with a set of keys in the event of an incident affecting the premises concerned.

keyhole the opening in a lock body or CASE in which is inserted the key. (See diagram 9 on p. 69)

keying the way in which the locking system of a building is arranged.

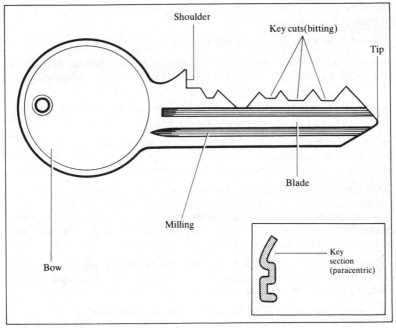

Diagram 5 *Cylinder or pin tumbler key*

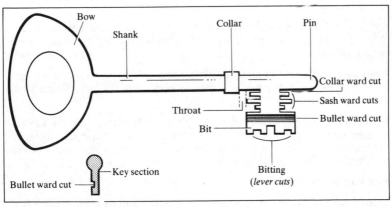

Diagram 6 *Lever lock key*

K

These may be INDIVIDUAL KEYING, KEYED-ALIKE, KEYED-DIFFERENT and MASTER KEYED systems.

key-in-the-knob lock see KNOBSET.

keyless combination lock a lock mechanism which operates by dialling a numerical code. The COMBINATION of such locks is easily altered by the owner using a CHANGE KEY (2).

keyless entry system an ACCESS CONTROL SYSTEM utilising DIGITAL LOCKS, ACCESS CARDS or equipment other than a key.

key-operated bolt a type of bolt which can be opened only with a key. Such bolts are often used as a SUPPLEMENTARY LOCKING DEVICE for metal windows. May also be known as a key-operated surface-mounted bolt.

key-operated mortice security bolt a simple rack-operated device installed in wooden doors as an alternative to conventional surface-fitted bolts. Also known as a rack bolt or mortice bolt. A similar (but shorter) device is used to secure wooden framed windows and is known as a window bolt. (See diagram 13 on p. 97)

keypad a method of entering data into an electronic system. Keypads are a frequent feature of ACCESS CONTROL SYSTEMS in which entering the correct sequence of numbers or letters allows operation of a lock or door.

keypoint a military and police term for a location which is deemed to be essential to the normal functioning of a community. Keypoints would normally include electricity generating stations, water plants, telecommunications equipment and broadcasting stations as well as structures such as bridges.

key safe a small wall SAFE used for secure storage of keys. It is often located outside buildings to permit access of fire or police personnel in an emergency.

key switch an electrical switch which can be operated only by a key. SHUNT SWITCHES and CONTROL KEY SWITCHES are the most common forms of key switch.

keyway the opening in a LOCK or CYLINDER, which is shaped to accept a key BIT or BLADE of the correct cross-section. (See diagrams 9 and 10 on pp. 69–70)

keyway blocking disc a steel disc or pin located deep within the KEYWAY of high security PIN TUMBLER LOCKS. The presence of the disc increases the number of theoretical COMBINATIONS, although it does not provide high security. This device is also known as a profile blocking disc.

kick bar see FOOT SWITCH.

kidnap and ransom a type of insurance pioneered by certain syndicates at Lloyd's of London. Essentially, the policy indemnifies the policyholder against the payment of ransom. There are a number of conditions, the most important of which is that the named individual is compelled to take advice on EXECUTIVE PROTECTION and act on any mandatory improvements suggested by the specialist company nominated by the underwriters of the policy. The other important condition is that the existence of the policy and the extent of cover must remain a secret.

kniferest a military term used to describe a wooden or steel structure used for temporary roadblocks or crowd control. It normally consists of a crossbar supported by two legs at each end. See also CHEVAL DE FRISE.

knobset a locking device in which the CYLINDER LOCK mechanism is contained inside the door knob. The lock body and BOLT mechanism is contained within the door, but this device is not treated as a MORTICE LOCK. Also known as a lock-in-knob or key-in-the-knob lock.

knock-out rods a method of protecting large areas of glass, particularly skylights, which consists of a series of microswitches held closed by a row of thin metal tubes or wooden battens. If any of the tubes or battens is dislodged by an intruder, the alarm is triggered. This type of sensor is also suitable for use in ducts, cable tunnels, crawlspaces and above false ceilings.

L

lacing see TUBE AND WIRE.

lag in closed-circuit television systems, the interval between the display of successive images on a MONITOR operated by a SEQUENTIAL SWITCHER.

laminate 1. a very thin layer of metal FOIL bonded to a non-conductive material that is electrified when an alarm is switched on. Any penetration of the foil results in the alarm being triggered
2. a double film of plastic material which, when heat-sealed, envelops totally an IDENTITY CARD both as protection against wear and as prevention against alteration.

laminated glass a combination of glass and plastic (the plastic often taking the form of a polycarbonate). This is used to give greater protection to windows which do not warrant the extra expense of BULLET-RESISTANT GLAZING or BANDIT-RESISTANT GLAZING. Laminated glass can also incorporate wired mesh to provide a degree of fire resistance. See also GEORGIAN GLASS.

laser protection system a BREAK-BEAM ALARM type of sensor (similar to a PHOTO-ELECTRIC SENSOR or INFRARED BEAM) where the light source is generated by a very low power laser. The use of a laser offers advantages over other light sources due to its greater range and coherent beam.

latch a door fastening device which uses a BOLT, usually a bevelled SPRINGBOLT. Latches normally keep doors closed, but not locked. (See diagram 9 on p. 69)

latch bolt the spring-loaded BOLT or live bolt in a lock or LATCH. (See diagram 9 on p. 69)

latch bolt hole the aperture in a STRIKING PLATE through which the LATCH BOLT protrudes when a door is closed. (See diagram 9 on p. 69)

latching (unit) a term or device used in connection with the extension of existing alarm installations when all circuits of a CONTROL PANEL are in use. Some alarm companies suggest that a latching unit can be used so that additional sensors can be connected to the latching unit rather than direct to the control panel. Thus, a single latching unit may 'control' a number of new sensors. Latching units are often fitted with indicating lights (usually LEDs) which are known as latching or WALKTEST LIGHTS, and when a system is UNSET it is possible to use the latching unit for walktesting the sensors. Care should be taken that the latching unit is connected to the control panel in such a way as it is not possible to SET the control panel while the latching unit is in a 'test' position.

LEAA (US Department of Justice) Law Enforcement Assistance Administration.

leaf a part of a pair of doors or shutters which comprise two or more hinged parts. For example, a double door has two leafs (or leaves), one active (or first-closing), the other passive.

leased line a communications circuit (normally rented from a telephone company) for use by the subscriber to transmit alarm signals or other data.

LED light-emitting diode: an electronic component which illuminates when energised. Frequently found in CONTROL PANELS and MIMIC PANELS where its low power consumption and long life is valued. LEDS have virtually superseded filament bulbs for this type of use.

letter bomb an IED, usually sent through the mail, consisting of a small explosive charge intended to detonate when opened.

lever a flat, spring-loaded, lock TUMBLER which pivots on a POST. (See diagram 7 on p. 67)

lever and warded mechanism an older type of lock which uses WARDS as well as LEVERS to provide a greater number of combinations.

lever handle a mechanism for operating a LATCH BOLT. Lever handles work by a simple press action and are preferred to rotating knobs. The lever handle is now mandated in many countries by legislation designed to help the handicapped or elderly. (See diagram 9 on p. 69)

lever (lock) key a key intended for use in a locking device fitted with a LEVER MECHANISM. (See diagram 6 on p. 61)

lever (lock) mechanism in a LEVER lock, the levers contain GATES which the correct key aligns to allow movement of the bolt. This is done by passing the BOLT STUMP (or fence) through the gates. (See diagram 7 on p. 67)

lever notching a term for the COMBINATION cuts on a LEVER LOCK KEY.

lever pivot the POST in a lock on which the LEVERS swing. (See diagram 7 on p. 67)

lever spring the spring which tensions the LEVER against the POST. (See diagram 7 on p. 67)

lie detector see POLYGRAPH.

limpet see SAFE COVER.

line amplifier an electronic device for boosting a closed-circuit television or alarm signal to increase the distances possible between location of cameras and monitors, or between alarm sensors and their control equipment.

line signalling system see DIRECT LINE SIGNALLING.

line supervision a method of providing high security for an alarm system communication circuit by introducing a continuous impedance or other coded signal into the circuit. If the line is cut or otherwise tampered with, this is sensed and an alarm will then be triggered at either end of the circuit.

lip the part of a STRIKE on which the LATCH BOLT rides. The lip is usually slightly curved. (See diagram 9 on p. 69)

live bolt see LATCH, LATCH BOLT.

LLTV Low-Light TV: a form of closed-circuit television which operates at lower than normal light levels. See also LOW-LIGHT TELEVISION CAMERA.

local alarm 1. a bell, siren or other SOUNDER fitted externally which is triggered when an ALARM CONDITION exists
2. any alarm system which is not connected to a CENTRAL STATION or other remote monitoring point.

local sounder see LOCAL ALARM (1).

local warning system see LOCAL ALARM (2).

Locard's Principle the basis of all scientific crime detection, also known as the Theory of Transference or the Theory of Interchange. The principle holds that two objects cannot come together without there being some exchange of materials. Thus, when an intruder enters a room, he picks up some part of the room contents (fibres from furnishings, blood from a corpse) and leaves behind fragments from his body and clothing (hair, skin cells, fingerprints, dust from shoes and clothing fibres).

lock a device usually but not always key-operated, having one or more BOLTS and used to secure a door, window or drawer. Diagrams 7–12 on pp. 67–71 indicate the main features of various types of locks and fittings.

lock case the metal box into which are fitted the component parts of a lock. Also known simply as a case. (See diagram 9 on p. 69)

lock cover a plain or unadorned ESCUTCHEON.

lock face see FOREND.

locking anchor point 1. the point at which a SUPPLEMENTARY LOCKING DEVICE is secured, especially in relation to securing windows
2. a type of locking plate for preventing the theft of items such as computers or typewriters.

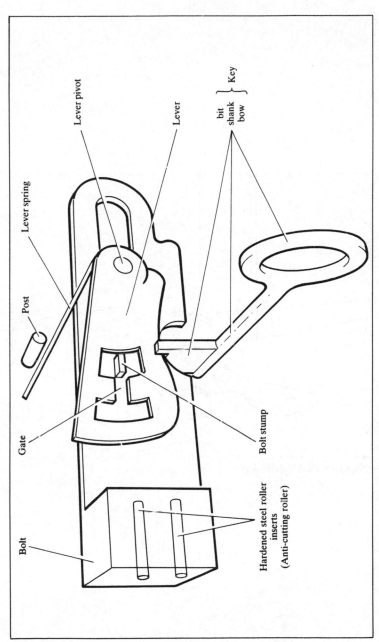

Lever spring

Lever pivot

Post

Lever

Gate

bit
shank } Key
bow

Bolt stump

Bolt

Hardened steel roller
inserts
(Anti-cutting roller)

Diagram 7 *Internal components of a lever lock*

Diagram 8 *Push or patio lock*

L

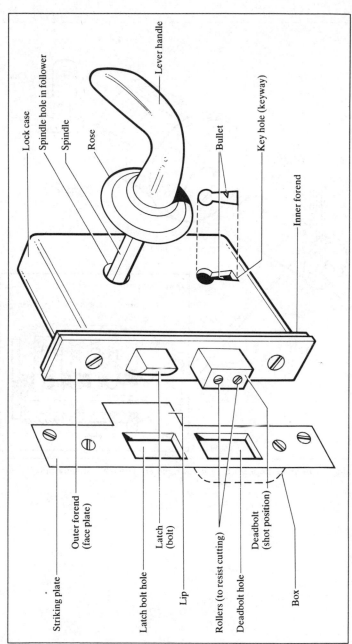

Diagram 9 *Mortice lever lock*

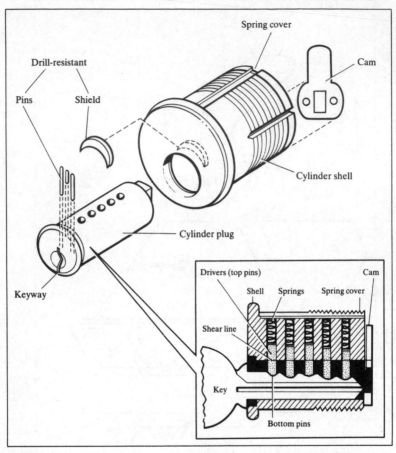

Diagram 10 *Internal components of a pin tumbler cylinder* (showing pins)

L

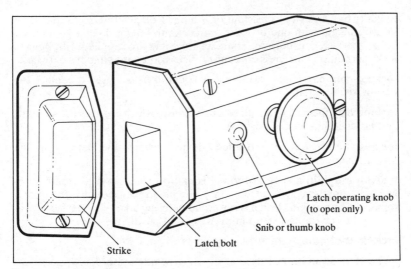

Latch operating knob
(to open only)

Snib or thumb knob

Strike

Latch bolt

Diagram 11 *Rim nightlatch*

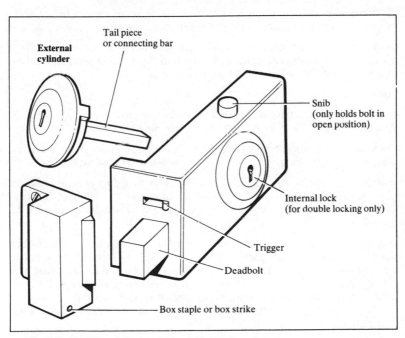

Tail piece
or connecting bar

**External
cylinder**

Snib
(only holds bolt in
open position)

Internal lock
(for double locking only)

Trigger

Deadbolt

Box staple or box strike

Diagram 12 *Rim automatic deadlock or deadlatch*

locking bar a device consisting of a hinged or pivoting metal bar used in conjunction with one or more PADLOCKS to secure double doors from either the inside or outside. Some locking bars are supplied with built-in lock mechanisms for greater security. See also CLOSE SHACKLE PADLOCK.

locking cam catch a SUPPLEMENTARY LOCKING DEVICE used to replace the CAM or FITCH CATCH.

locking dog the part of a PADLOCK mechanism which engages the SHACKLE and holds it in a locked position.

locking latch any LATCH-fitted device where the latch can be DEADLOCKED.

locking stay pin a SUPPLEMENTARY LOCKING DEVICE used to secure casement WINDOW STAYS. It consists of a simple screw cap which fits over a replacement stay pin. When screwed home using a key-like device, it will secure the stay on the stay pin in the closed position.

lock-in-knob (mechanism) see KNOBSET.

lockout 1. an industrial dispute in which workers are denied access to their workplace
2. any situation in which a key or cylinder is inoperable
3. the deliberate act of applying a padlock or other device to a circuit breaker or piece of machinery as a safety measure.

lockout key a key made in two pieces. The first is inserted into a lock to prevent other keys operating; the second piece being used to remove the first when the maintenance activity is over.

lock pick see PICK.

lock shoot the distance or depth which a bolt moves when it is operated. Usually refers to a LIVE or LATCH BOLT. See also BOLT THROW.

lock switch a modified lock, fitted with electrical contacts or a micro-switch, normally used as a SHUNT LOCK. See also KEY SWITCH.

lodge a GATEHOUSE or GUARDHOUSE often also providing accommodation for a security guard or gatekeeper. Also known as a gatelodge. In Scotland, this term is often used commonly as a synonym for gatehouse or guardhouse.

log a written record of a tour of duty or of activity carried out in that tour. For example, among the books kept in a GATEHOUSE or security control room would be a Visitor's Log, a Key Issue and Receipt Log and so on. The term is also used to refer to an automatically printed record of events produced by a computer-controlled security system, sometimes known as an 'event printer' or 'data logger'. Logs are often also known as registers.

loiding the action of bypassing the PIN TUMBLER MECHANISM of a lock and moving the bolt with a thin but rigid piece of plastic or steel. The term is derived from 'celluloid(ing)' but today, credit cards would appear to be more frequently used – at least on television. Also known in the US as shimming.

long firm fraud a common fraud where a criminal team sets up a 'business', builds up a relationship with suppliers and then obtains large quantities of goods on credit. The goods are speedily sold (often at a discount) and the 'business' closes leaving substantial debts.

Location	Light level (Lux)
Outdoors, full moon	.001
Outdoors, twilight	5–10
Outdoors, overcast day	1,000
Bright sunlight	10,000
Car parks	10–20
Site entrances	50–100
Corridors and stairs	100
Security gatehouses	200
Drawing offices	750

Table 2   *Typical light levels*

loop a circuit that starts and ends at the same location.

loop alarm (system) an electronic system used for the protection of higher value goods like stereo equipment, cameras and the like in retail establishments. The system consists of lengths of cable joined together by jack plugs and sockets. The loop is plugged into a CONTROL UNIT which is fitted with a SOUNDER and a KEY SWITCH. When the loop is passed through an item, any attempt to break or cut the wire or unplug it will result in an alarm being triggered.

looping 1. a method of wiring sensors or other devices to share a single coaxial or other circuit
2. a closed-circuit television switching device which permits video inputs to pass through the switcher and on to a monitor or other device without terminating at the switcher.

loss prevention a term which is often used as a synonym for ASSETS PROTECTION. The phrase is becoming more commonly used in major companies as a name for the department which controls safety and security.

Loss Prevention Council a UK-based insurance-sponsored body involved in the promotion of fire prevention and the reduction of fire losses. Actively involved in arson prevention through better security, as well as listing and approval of fire fighting equipment.

L

louvre window a window consisting of a number of wooden or glass slats set in a frame and capable of being adjusted to provide ventilation. Such windows provide very poor security.

low-light television camera a closed-circuit television camera which uses special circuitry to produce an acceptable image under light conditions so poor that a normal camera would not produce an acceptable image. See also NEWVICON, NEWVICON INTENSIFIED TUBE, SIT.

lumen see LUX.

lux the SI measure of light intensity used in calculations for both closed-circuit television systems and security lighting. The lux equals one lumen per square metre (the older FOOT CANDLE is 10.764 lux). Some examples of light levels relevant to security work are shown in table 2 on p. 73. See also SECURITY LIGHTING.

L

magnetic buried line sensor a form of PERIMETER PROTECTION where a loop of cable is installed up to 200 mm below the surface of the ground. When a vehicle passes over the loop, the magnetic field alters, triggering an alarm. A similar system is used to actuate car park barriers and in this case, the equipment is usually described as a detection loop.

magnetic card a plastic card manufactured with some form of magnetic content which can be encoded with data and used as an ACCESS CARD or as an ATM card. The term 'magnetically encoded card' is also used for this type of card. Examples of this type of card are the MAGNETIC SPOT CARD, MAGNETIC STRIP(E) CARD and WEIGAND CARD.

magnetic contact a magnetically operated switch, typically used on doors and windows. Operation is facilitated by a permanent magnet fixed to the door or window while the 'switch' is fixed to the frame or non-moving part. A reed switch is a magnetic switch which consists of two thin steel leaves or 'reeds' mounted in a glass cylinder. When the cylinder is brought close to a magnet (as when a door is closed) the reeds flex making contact and completing a normally open circuit.

magnetic lock a door-securing device in which an electromagnet takes the place of the bolt and locking mechanism. A steel plate is securely fastened to the top side of the door and the electromagnet is fitted to the door frame. The lock is operated by energising or de-energising the magnet which can be done by an ACCESS CONTROL SYSTEM or a push-button. This type of lock is extremely strong and is especially suitable for high security applications given its extreme reliability. Care must be taken to specify whether the device should FAIL-SECURE and appropriate measures should be taken with respect to the provision of power supplies. Magnetic locks are also known as electromagnetic locks.

magnetic sensor see MAGNETIC CONTACT.

magnetic spot card an ACCESS CARD where the information is coded in 'spots' on a barium ferrite core rather than on a magnetic stripe. (See table 1 on p. 17)

magnetic strip(e) card an ACCESS CARD where the information is coded in the form of a strip or stripe of magnetic tape bonded to the card. (See table 1 on p. 17)

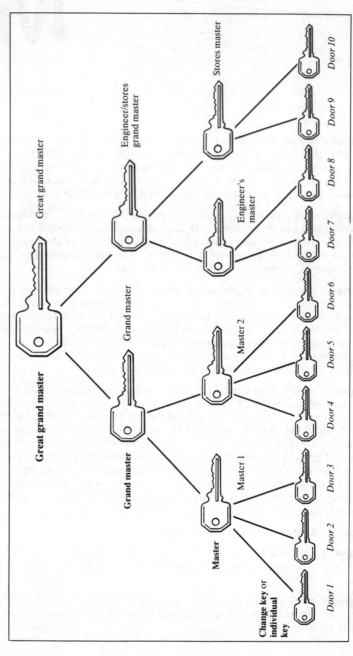

Table 3 Simplified master key system (four levels)

magnetic switch see MAGNETIC CONTACT.

maison key system a keying system in which one or more cylinders are operated by every key in the system. Commonly used in apartment complexes where the main entrance door locks can be operated by each individual apartment key.

malachite green a moisture-activated VISIBLE STAINING COMPOUND which produces a vivid green colouration on skin contact.

Manbarrier a proprietary name for the US GPBTO and the company which originally manufactured it.

manipulation the operating of a locking device by any means other than that intended.

manned post a security post or location which is staffed by a guard or officer. Such posts are also called static posts.

mantrap 1. an illegal device intended to harm intruders or trespassers. Mantraps are usually mechanically-sprung devices (similar to animal traps) in which substantial jaws are released when a foot is placed on the base of the trap
2. in US security terms, the space between two INTERLOCKED DOORS (known in the UK as an airlock).

manual (active) switcher a closed-circuit television system component which selects which image is to be displayed on a MONITOR.

manual McCulloh (system) see MCCULLOH-TYPE SYSTEM.

Mareva Injunction an English High Court order which freezes the assets of a defendant preventing their dispersal prior to a civil suit. Often used in cases of product counterfeiting, parallel trading, fraud and embezzlement.

marine fraud a range of frauds which have increased greatly in recent years, mainly involving the issue of forged or falsified bills of lading and letters of credit, or taking out insurance (and later claiming) on worthless or non-existent cargo. This is a major growth area in crime and one in which private security can be closely involved because of the difficulty of involving law enforcement agencies where national jurisdiction is unclear.

master code card the ACCESS CARD equivalent of a combination CHANGE KEY which is used to program(me) the appropriate codes into a CARD READER.

master key a key which operates all the MASTER-KEYED locks or CYLINDERS in a group; each lock or cylinder is also operated by its own CHANGE KEY.

master keyed (system) a method of arranging locks so that they operate on a hierarchical basis as illustrated in table 3 on p. 76.

Mastiff System a proprietary form of HANDS-FREE ACCESS CONTROL.

matching network the electronic circuitry used to enable an alarm monitoring line to be able to transmit audio signals. Such a circuit would be used in an AUDIO MONITORING system.

mat switch see PRESSURE PAD.

MBF a colour-corrected mercury discharge lamp with similar characteristics to the MBI lamp, but which is less economic to run.

MBI a metal halide lamp with a lower level of output per watt than SOX or SON lamps but having excellent colour-rendering characteristics.

McCullogh see MCCULLOH.

McCulloh Circuit an electronic device which enables a number of alarm systems to share a common transmission circuit. Each alarm system transmits a coded signal so that it can be identified at the CENTRAL STATION. Also known as a McCulloh Loop.

McCulloh Loop see MCCULLOH CIRCUIT.

McCulloh Receiver a device located at a CENTRAL STATION which decodes signals from a MCCULLOH CIRCUIT either on a VDU or a printer.

McCulloh Transmitter a device which transmits the coded signals over a MCCULLOH CIRCUIT. Recent developments have given what is a relatively old system a new lease of life by allowing a variety of signals to be transmitted, thus a conventional intruder alarm signal could be simply and economically modified to send plant, fire and intruder alarms on the same circuit using the same transmitter and receiver equipment.

McCulloh-type system an electronic circuit which enables an alarm device such as a SENSOR to trigger an alarm despite the presence of a short-circuit. This type of device can be manually selected as in the manual McCulloh system or automatic as in the automatic McCulloh system.

MCF a tubular fluorescent lamp of similar characteristics (relatively long life, low cost) to the conventional domestic or industrial 'tube'.

mechanical contact a type of SENSOR based on a microswitch or SPRING CONTACT.

mechanical door lock a non-powered locking device. Sometimes used to describe a conventional key and lock system, more often to refer to a DIGITAL LOCK.

mechanical lock a term occasionally used to refer to a lock which contains movable components such as springs, PINS, or LEVERS operating on mechanical principles of force, motion and friction.

mechanical pick see PICK GUN.

mechanical vibro-contact a now obsolescent type of SENSOR consisting

M

of a leaf spring and weight (fixed to the surface to be protected), which are attached to a simple contact switch. Movement of the protected object results in vibration of the spring which results in the switch being operated by the weight triggering an alarm.

mercury-drop contact switch see TILT SWITCH.

metal detector a fixed or hand-held device designed to detect and indicate the presence of metallic objects. Metal detectors are used at airports to prevent the carrying of firearms or explosive devices and are also used in major industrial plants to deter theft of tools etc.

metallic foil see FOIL.

microchannel plate intensifier a closed-circuit television camera component which increases the efficiency of a NEWVICON enabling it to operate at very low light levels.

microphonic device see AUDIO DETECTION.

microphonic fence protection a PERIMETER PROTECTION SYSTEM based on ELECTRET CABLE.

microwave motion sensor a MICROWAVE SENSOR operating on the principle of DOPPLER SHIFT.

microwave sensor an intruder SENSOR which detects movement of persons or objects within an emitted field. Microwave sensors can be classified into three groups: monostatic, using the DOPPLER SHIFT; bistatic, using the BREAK-BEAM ALARM principle, and TERRAIN-FOLLOWING SENSORS, using a line-of-sight method which resembles the bistatic type of sensor. Microwave sensors are the most commonly used form of AREA PROTECTION and although a later generation of PASSIVE SENSORS are now being increasingly used, when correctly installed in appropriate locations they remain very cost-effective.

MIlSec Member of the Institute of International Security see IISEC.

milling the pattern of grooves machined along the length of a CYLINDER LOCK key BLADE to allow its entry into the KEYWAY. (See diagram 5 on p. 61)

mimic panel a display, normally incorporating a map or schematic representation of a site, fitted with LEDs or bulbs which illuminate when an alarm is triggered to indicate the location of a problem.
   A synoptic map closely resembles a mimic panel but this term is normally used only to refer to a map or panel which encompasses an entire site or building.

MIPI Member of the Institute of Professional Investigators see IPI.

MLA Master Locksmiths' Association (UK).

mobile annunciator an alarm control or display panel which can be mounted in a vehicle.

mode the condition in which an alarm system finds itself, for example, ACTIVE MODE, ALARM MODE, FAULT mode or isolated mode.

modulated photoelectric sensor a PHOTOELECTRIC SENSOR which uses a modulated beam (normally in the INFRARED spectrum) to detect an intruder. Such devices cannot be overcome by using another infrared source as they respond only to the specific modulations of the original beam.

moisture detector a sensor which triggers an alarm in the presence of a predetermined quantity of water or moisture.

money clip a fitting in a cashier's drawer equipped with a simple spring-type switch connected to a BANDIT ALARM and usually supplied with BAIT MONEY.

money laundering the conversion of cash obtained through illicit means into usable and untraceable funds.

money trap see MONEY CLIP.

monitor a modified television set which displays the images produced by CCTV cameras.

monitoring station a location, remote from the PROTECTED AREA, at which security guards or operators monitor ALARM RECEIVERS or ANNUNCIATORS.

mortice bolt see RACK BOLT.

mortice cylinder lock a MORTICE LOCK fitted with a PIN TUMBLER MECHANISM.

mortice deadlock a DEADLOCK fitted within a door or window STILE.

mortice lever lock a MORTICE LOCK fitted with a LEVER MECHANISM.

mortice lock a lock mounted within the body of a door or window frame. (See diagram 9 on p. 69)

mortice sashlock a MORTICE LOCK with both LATCH BOLTS and DEADBOLTS especially designed and manufactured to fit narrow door STILES.

mortice shield a device in the form of metal plates, fitted to strengthen a door STILE which may have been weakened by the removal of wood to install a MORTICE LOCK.

mortise lock see MORTICE LOCK.

motion detection the use of any form of MOTION SENSOR to trigger an alarm.

motion detector see MOTION SENSOR.

motion sensor 1. a SENSOR which responds to movement. Motion sensors may be part of a conventional alarm system, such as a MICROWAVE SENSOR, or may provide a trigger mechanism for other devices such as automatic doors

2. in CCTV systems, motion sensors are electronic, programmable devices which enable an operator to set up a grid pattern on a MONITOR. Any change within this pre-set pattern caused by movement of people or vehicles results in an alarm being triggered which will attract the attention of the operator, can operate a remote alarm and start a video recorder.

Movealarm a proprietary name for one of the earliest MOTION SENSORS.

multicoupler an impedance-matching device which permits several receivers to share a single aerial.

multiple latch a simple mechanical device, usually consisting of three or four toothed LATCHES, used with a PADLOCK to secure gates in fences.

multiplex signalling (system) an electronic device which enables a single circuit to carry more than one signal. See also ABC, BASE 10, FREQUENCY DIVISION MULTIPLEX, TIME DIVISION MULTIPLEXING, VOICE GRADE MULTIPLEX SIGNAL.

multipoint locking handle a simple locking device for windows where a handle is connected to a rod which secures two or more catches. Similar to an ESPAGNOLETTE.

multipoint locking system a type of locking system which is mechanically connected to more than one BOLT or catch. Such devices often SHOOT bolts into the top and bottom of the door frame as well as the sides.

mushroom (pins) the top or DRIVER PINS in a PIN TUMBLER LOCK are often shaped like inverted mushrooms. This increases the level of security by making PICKING more difficult. See also SPOOL PIN, HUCK PIN.

mutual aid a system where a number of companies combine to provide additional security for their premises. This could either be by jointly hiring a guard or patrol service or forming a TELECONTACT network. Mutual aid schemes can also be formed to provide assistance following fires or other disasters.

N

NACA National Armored Car Association (US).

nail bomb an IED which normally consists of a bundle of six-inch nails taped around a stick of commercial explosive.

NAR Nuisance Alarm Rate see NUISANCE ALARM.

narrow stile lock an upright MORTICE LOCK with a shallow body to fit doors with thin jambs or stiles. Also known as a SASH LOCK.

NASSD National Association of School Security Directors (US).

NBFAA National Burglar and Fire Alarm Association (US).

NCISS National Council of Investigation and Security Services (US).

NCPC National Crime Prevention Coalition (US).

NCPI National Crime Prevention Institute (US).

negative vetting see NV.

negative image a closed-circuit television system fault resulting from polarity reversal during signal transmission. The image has the light and dark areas reversed.

neighbourhood watch a US concept, much-adopted worldwide, where residents of a locale combine under police control to reduce crime by means of observation and reporting, property marking and displaying posters.

Newvicon a proprietary name for a closed-circuit television camera tube and associated circuitry. The Newvicon offers greater sensitivity and longer life than the conventional Vidicon tube and will operate satisfactorily over a greater range of light levels down to 5 LUX. The Ultricon tube has very similar characteristics to the Newvicon. See also NEWVICON INTENSIFIED TUBE.

Newvicon Intensified Tube an improved version of the NEWVICON tube which like the SIT will operate at light levels from 2 LUX down to 0.001 lux.

NFPA National Fire Protection Association (US).

Ni-Cad battery a nickel cadmium cell or battery used extensively in security applications where reliable, low maintenance, stand-by power is needed.

nightlatch a low-security locking device still found in many homes. Usually of the RIM LOCK type, with a PIN TUMBLER MECHANISM, such a lock is fitted with a button (or 'snib') which can DEADLOCK the bolt from the inside (when the premises are occupied). The button can also hold the bolt in the withdrawn position when desired. (See diagram 11 on p. 71)

999 system an INTRUDER ALARM SYSTEM connected to a TAPE DIALLER configured to call the (UK) police emergency number if an alarm is triggered.

noise detector see AUDIO DETECTION, AUDIO MONITORING.

non-coherent fibre bundle see FIBRE OPTICS.

non-drying paint see ANTI-CLIMB PAINT.

non-linear junction detector an electronic device used in the detection of BUGS which operates by the emission of high frequency signal (915 MHz) and then waits to receive the harmonics of that signal back from the non-linear components contained in the bug's own circuitry.

normal sensor a sensor which triggers an alarm only when the control panel is SET. The opposite would be any sensor connected to a PERMANENT CIRCUIT.

normal vetting see NV.

notching see BITTING.

Notecalm Alarm see SELF-CONTAINED INTRUDER ALARM.

NSCIA National Supervisory Council for Intruder Alarms: a UK body which approves alarm installers and issues alarm installation approval certificates.

NSIA National Security Industrial Association (US).

nuisance alarm an unwanted alarm normally resulting from activity of people, animals, birds, vegetation or climate which simulates the stimulus of an intruder. See also FALSE ALARM.

NV Normal Vetting, (sometimes) Negative Vetting, a form of BACK-GROUND SCREENING. The term normally refers to governmental clearances done for less sensitive posts. See also PV.

O

object protection see SPOT PROTECTION.

off-line a device is said to be off-line when it is not connected to a central computer or processing unit. When the device is connected, it is 'on-line'.

off-line card reader the card reader component of an ACCESS CONTROL SYSTEM which is equipped with its own memory and power supply.

offset 1. a term used to describe the distance on the ground (when measured from a transmitter or receiver) along a beam path which would permit an intruder to move without being detected
2. the overlapping of a beam span by staggering transmitters and receivers to prevent beam gaps which may be used by intruders.

on-line a device is said to be 'on-line' when it is connected to a central computer or data link system.

on-line card reader a CARD READER (usually within an ACCESS CONTROL SYSTEM) which is permanently connected to a central computer.

open the state of an INTRUDER ALARM SYSTEM in which an ALARM CONDITION cannot be signalled.

open-circuit alarm system a system in which sensors or other devices are connected in parallel.

open-circuit device a SENSOR or other detection device which is designed to trigger an ALARM CONDITION when an open circuit is closed.

open detection circuit a detection circuit which is in ALARM CONDITION when it is closed.

opening signal an alarm signal transmitted to a CENTRAL STATION at a predetermined time each day. Opening signals are agreed between the two parties and any deviation from an agreed time frame will result in the central station's dispatching the police. Closing signals are similarly organised.

operational security security management techniques as they are applied to the control of property, processes and people by administrative methods.

optical density card an ACCESS CARD, sometimes known as an INFRARED (access) card, made of a translucent material in which are imbedded areas of varying but controlled density. Each variation represents a numerical value and such cards offer very high security as they are difficult to read or copy. (See table 1 on p. 17)

optical fibre fence a fence ENHANCED with some type of FIBRE OPTIC equipment.

outer forend see FOREND.

out-of-band-signalling a method of DIRECT LINE SIGNALLING using a telephone line where the alarm impulses are transmitted outside the normal speech frequency bands (300–3,000 Hertz).

outside perimeter a line of protection close to but external to a perimeter marker such as a wall or fence.

P

PA button personal attack button see ATTACK BUTTON.

PACE Police and Criminal Evidence Act 1984: a piece of UK legislation which is of interest to the security professional in that it provides the general citizen's (and therefore the security officer's) power of arrest (previously provided by the Criminal Law Act 1967).

The Act and its associated Code of Practice also mandate the ways in which statements should be taken and cautions administered to suspected persons. This codifies for the first time instructions to the police and 'other persons charged with the investigation of offences' originally contained in the Judges Rules.

While the Code of Practice does not appear to apply to the security officer in the private sector, it would seem to be sensible for those practising security in the UK to follow it whenever possible.

pad see PRESSURE PAD.

padbolt a TOWER BOLT with the facility for locking the BOLT in the SHOT position using a PADLOCK.

PA device personal attack device see PORTABLE DURESS ALARM.

padlock a portable device, usually key operated, with a hinged or sliding SHACKLE to pass through an eye or staple or through the item to be safeguarded. See also HASP AND STAPLE.

pale see PALISADE FENCE.

paling see PALISADE FENCE.

palisade fence a type of fencing constructed from vertical steel sections known as pales or palings held by bolts to lateral rails. The tops of the vertical sections can be shaped to discourage climbing. When palisade fencing is specified for a security application, the pales should be corrugated.

pan and tilt the facility provided by a motorised device to enable a closed-circuit television camera to be remotely aimed to any point on a 360 degree circle and to move to about 15 degrees above and below the horizon.

panel see CONTROL PANEL.

panic alarm a feature found in domestic INTRUDER ALARM SYSTEMS. Like the RAID or BANDIT ALARM, panic alarms are wired into a PERMANENT CIRCUIT (the raid or panic circuit) which is always active even when the main alarm system is off. The primary difference between the two types of system is that a raid alarm is always silent (sounding no local alarm) while the panic alarm will always immediately trigger an external bell even where the alarm system is connected to a CENTRAL STATION.

panic bar a mechanical fitment which is installed on fire exits, particularly those in public places. The door is secured by the device with a top and bottom BOLT, but pressure on the bar which is mounted at waist height will withdraw the bolts and allow the door to open easily. This type of fitment is also known as a CRASH BAR or exit device. It is relatively simple to fit CONTACTS to such a door or to fit an alarmed exit device which contains a battery and SOUNDER. Any attempt to open the door will result in an alarm.

panic bolts see PANIC BAR.

panic button see PANIC ALARM.

parabolic microphone a sensitive microphone fitted with a concave reflector often used as an EAVESDROPPING DEVICE.

parabolic mirror a concave mirror used in conjunction with PASSIVE INFRARED SENSORS to define areas to be covered by the system.

paracentric a term used to describe a key or KEYWAY where one or more WARDS on each side project beyond the vertical centre line of the key or keyway. Most PIN TUMBLER keys and keyways are paracentric. (See diagram 5 on p. 61)

para-rosaline sulphate a moisture-activated, VISIBLE STAINING COMPOUND producing a distinctive purple colouration on skin contact.

passback a method of gaining entry to a building or facility where an individual in possession of an ACCESS CARD enters the building and then passes the card through a fence or out of a window to an accomplice. A system not fitted with an ANTI-PASSBACK CIRCUIT will fail to detect this subterfuge.

passive infrared (sensor) a device capable of detecting changes in emitted heat within its pre-set detection range and inside its field of view. The SENSOR does not transmit any energy, and is thus termed 'passive'. PIRs offer many advantages over other types of sensor, principally low current drain and reliability. They are also free from false alarms created by electromagnetic emissions in the vicinity of the sensor.

passive microwave reflector a method of extending the coverage of an active MICROWAVE SENSOR system. A flat, metallic reflector is installed to 'bounce' microwave emissions around the area to be protected, covering uneven terrain or tight spaces and joining adjacent sectors. The unit is said to be 'passive' as it is not powered.

passive sensor any type of SENSOR which does not generate or transmit any energy but instead detects changes in the surrounding levels of natural radiation, noise or air pressure.

passive ultrasonic detector a detector which responds to certain sound frequencies, which can be pre-set so that such a system can be 'tuned' to eliminate false alarms.

password 1. (originally) a military convention; a phrase or sentence used to confirm the identity of a party being challenged
2. now generally taken to be an identification word, phrase or number sequence used to gain access to a computer system.

patio lock see PUSH LOCK.

patrol service a service offered by CONTRACT SECURITY companies where a security officer makes an agreed number of visits to a location each day. Less effective than having a permanent guard or MANNED POST, but considerably less expensive.

pattern locator the part of a VOLUMETRIC SENSOR which permits the user or installer to define the boundaries of the sensor's protection pattern.

PD Probability of Detection.

peep hole see DOOR VIEWER.

percentage supervision a US term applied to a detection device or SENSOR which has a pre-set or predetermined reference value built into its circuitry. When there is a change in activity within the PROTECTED AREA, the sensor compares the value of the change with its reference value. The ratio of the change required to trigger an alarm, expressed as a percentage, is known as the percentage supervision.

perimeter the outer limits of a site, building or facility. In general terms, the legal boundary of a piece of land.

perimeter barrier any physical means used to delineate the legal or actual PERIMETER of a site.

perimeter detection (system) any device or collection of devices intended to trigger an alarm if an intrusion takes place through, under or over a PERIMETER barrier.

perimeter protection the varying methods of preventing unauthorised ingress through, under or over a PERIMETER. The simplest and most common example would be a fence. A more sophisticated system would incorporate a fence with some lighting, then with some form of detection system, and finally two fences, both equipped with detection systems. (See diagram 2 on p. 29)

perimeter protection system a combination of equipment used to prevent access to a given area. Normally taken to mean an ENHANCED fence together with some form of monitoring system. An example would be a CHAIN LINK FENCE, topped with BARBED TAPE and fitted with a fence protection system.

permanent circuit a circuit which is capable of alarm transmission regardless of the alarm system's operational MODE. See also PANIC ALARM.

personal attack alarm see PORTABLE DURESS ALARM (1).

personal attack button see ATTACK BUTTON.

personnel attack button see ATTACK BUTTON.

pet mat a type of PRESSURE PAD specially designed to be installed in houses where there is a dog or cat. The mat will not trigger an alarm if walked on by the pet as it is less sensitive than normal pressure-pads.

phone entry system see DOOR PHONE.

phone freak an individual who habitually obtains free telephone calls by manipulation of the telephone system. The term is usually applied to those who use a device called a 'black box' (to generate tones which bypass the telephone service's charging equipment) to make free international and long-distance calls.

phone phreak see PHONE FREAK.

photo-badge any IDENTITY CARD bearing a photograph of the holder.

photo-badge exchange a system whereby an individual is issued with an IDENTITY CARD bearing his photograph. On arrival at work, this card is handed over in exchange for a distinctive badge (which may be coloured differently each day) or TOKEN. The system is effective as it enables a detailed comparison of the identity card and its holder to be made each day and at the same time allows for an exact check to be made on workers present. If the identity card is lost, it is of little value. A similar system may be operated at some sites where a VISITOR BADGE is only issued in exchange for a driving licence or some other document. This ensures that visitor badges are returned.

photocell a device which responds to changes in light levels. Principally used as the passive or receiving component in a PHOTOELECTRIC SENSOR system. Also frequently used to switch on external lighting at dusk.

photoelectric sensor (system) a SENSOR system which functions when a beam of light aimed at a PHOTOCELL is interrupted by the body of an intruder. The beam may be visible light or an invisible light such as INFRARED.

photographic surveillance camera see BANK CAMERA.

photo identification (card) a BADGE or IDENTITY CARD bearing a photograph of the holder, together with other relevant information.

physical barrier any physical defence against intrusion into a PROTECTED AREA including not only man-made defences such as walls, fences or moats but also natural features like mountains, rivers, lakes, marshes, cliffs and even the sea.

physical security the protection of property and staff by the use of manpower and physical means including ALARM SYSTEMS, PHYSICAL BARRIERS and SURVEILLANCE equipment. The term is often used to distinguish between this type of security and security based on investigation, audit and management systems.

pick a tool or instrument designed and used for the MANIPULATION of a lock mechanism without damage to the lock.

pick gun a lock-manipulating device using vibratory principles, variously known as a picking pistol, pick gun or mechanical pick.

picking the MANIPULATION of a key-operated lock mechanism through the KEYWAY without the use of a key, but using a PICK or other tools. Picking does not normally involve damage to the lock. Also sometimes called mechanical picking.

picking pistol see PICK GUN.

PID passive infrared detector see PASSIVE INFRARED (SENSOR).

piezo-audio indicator a small audio alerting device which generates a high-pitched tone when energised; used extensively in CONTROL PANELS.

piezo-electric detector a type of SENSOR, sometimes classed as a shock sensor, which can be directly mounted on the object being protected. The piezo-electric element generates an electric current when stressed and this phenomenon is used, for example, in a type of GLASS BREAK DETECTOR. This type of sensor is also used to protect safes.

piggyback 1. (verb) to gain entry to a secure location by closely following an individual who has a key or other means of access. Known in the US as tailgating
2. (noun) any system of signal transmission where two or more messages are sent together.

PIN Personal Identification Number see ATM.

pin the tip of a lever lock key. (See diagram 6 on p. 61)

pinned hinge a simple mechanical device resembling a HINGE BOLT.

pins see PIN TUMBLER MECHANISM.

pin tumbler cylinder lock see PIN TUMBLER MECHANISM.

pin tumbler lock see PIN TUMBLER MECHANISM.

pin tumbler (mechanism) a type of CYLINDER LOCK which has a rotatable PLUG connected to a BOLT. The plug is prevented from turning by a set of pin tumblers in the form of spring-loaded pins. These consist of two sets – the top pins or drivers (which are often hardened to be more resistant to drilling or MANIPULATION), and the bottom pins. When the correct key is inserted in the plug, the pins are raised so that the joint between the top and bottom pins exactly coincides with the SHEAR LINE between the plug and the cylinder body, thus allowing the plug to be turned to operate the lock. Most cylinders contain at least five sets of pin tumblers. (See diagram 10 on p. 70)

P

pipe bomb a type of IED consisting of a piece of steel pipe filled with commercial explosive, black powder or improvised explosive mixture. This type of device is normally detonated by a simple timing device or a mechanical actuator such as a TILT SWITCH.

pipe key a UK term for BARREL KEY.

PIR see PASSIVE INFRARED (SENSOR).

PISC Petroleum Industry Security Council (US).

pivot a stump or POST mounted on a lock CASE to act as a pivot for levers. (See diagram 7 on p. 67)

plate glass see ANNEALED GLASS.

plug the part of a CYLINDER LOCK into which the key is inserted and which turns with the key. (See diagram 10 on p. 70)

pocket alarm see PORTABLE DURESS ALARM (1).

pocket book a notebook issued to security personnel for the recording of matters of importance arising during the course of their duty.

point detection see POINT PROTECTION.

point protection the use of a system of SENSORS to protect an object or item such as a single door or safe. The opposite of VOLUMETRIC PROTECTION. Also known as spot or object protection. (See diagram 2 on p. 29)

polarity reversal circuit a very popular method of ALARM TRANSMISSION until the recent developments in microchip technology. The system uses a DC signal, normally at 24 volts, which is continuous; in the event of an alarm being triggered, the CONTROL PANEL transmitter reverses the polarity of the signal. The continuous signal provides good line monitoring so that any break is immediately apparent. Polarity reversal signalling is still used in many fire alarm systems.

police station unit see HOME OFFICE PANEL.

polling period see SCAN PERIOD.

Poll Start see ABC.

polygraph a device sometimes known as the lie detector. Its proponents hold that in the hands of a skilled operator it is capable of detecting stress at various levels and therefore enables the operator to comment on the truthfulness of a statement. The device measures and records blood pressure, pulse, respiration and the amount of perspiration exuded by the fingers. There is a substantial disagreement within the security and law enforcement community with respect to the validity of tests carried out on the polygraph. While the debate continues, many US corporations require staff to take polygraph examinations upon appointment and even in some cases annually. See also PSYCHOLOGICAL STRESS ANALYSER.

portable detection system a complete detection system capable of being transported from one location to another and quickly set up to provide an INTRUDER ALARM SYSTEM in, for example, a hotel room or exhibition area.

portable duress alarm 1. a small electrical or gas powered noise-maker which is often carried by women. Also known as a pocket alarm, personal attack device, personal attack alarm, panic alarm or rape alarm
2. an electronic device which transmits a radio signal to an alarm system CONTROL PANEL. This unit functions exactly like a PANIC ALARM but is fully portable.

portable intrusion sensor any type of intrusion detection system which can be taken from place to place and put into service within a short time. Such units are often self-contained in a single weatherproof enclosure and, if they are triggered, transmit a radio signal to a special receiver.

portable road block see DRAGON'S TEETH.

ported coax a BURIED DETECTION SYSTEM, resembling the E-FIELD SYSTEM. Special 'leaky' coaxial cable is buried up to 20 cm below the surface. One cable radiates an electromagnetic field and the other detects the field. If the field is broken or modified due to the presence of an intruder this is detected and the alarm triggered.

positive vetting see PV.

post a stump mounted on a lock CASE to act as a holder or bearing for springs or as a guide for other component parts. (See diagram 7 on p. 67)

powder-proof (lock) a type of SAFE LOCK, now obsolete, which supposedly prevented the use of gun-powder as a method of blowing the safe open. (Sometimes referred to as 'patent powder-proof'.)

pressure differential system a comparatively rare method of protecting a VAULT or similar location where fans are used to maintain air pressure within the vault higher than the ambient. Any drop in pressure (as would result from tunnelling) results in a drop in pressure which is recorded by a simple barometer-like instrument which then triggers an alarm. Although this type of system is obsolescent, when it is described, it is often referred to as a 'pressure differential system (air)' to distinguish it from the liquid-based underground system. The essential element of this type of system is also known as a PRESSURE SENSOR.

pressure differentiating system a BURIED LINE INTRUSION DETECTOR system using a BALANCED PRESSURE SENSOR and thus more commonly known as a balanced hydraulic detection system. Also known as a 'differential pressure system (liquid)' to distinguish it from the air pressure-based system.

pressure mat see PRESSURE PAD.

pressure pad a thin plastic or rubber pad or mat, usually consisting of two strips of conductive foil separated by a non-conductive layer. If pressure is applied to the surface of the mat, the foil faces come into contact completing the circuit. Some pressure pads contain a sensor which operates on the CAPACITANCE principle and these are much more sensitive. Pressure pads are usually fitted under carpets in front of doorways, beneath windows or at the top and bottom of stairways. See also PET MAT.

pressure-sensitive perimeter alarm (system) see BALANCED PRESSURE SENSOR.

pressure sensor a type of sensor which responds to changes in air pressure. Normally only found in high security locations with strictly controlled conditions as for example, in a bank vault. See also PRESSURE DIFFERENTIAL SYSTEM.

primary locking mechanism the modern lock, especially in high security areas such as VAULTS, sometimes uses two separate locking mechanisms, operated by the same key. One lock of this type combines SIDEBAR and PIN TUMBLER MECHANISMS. In this type, the key not only has to raise the pins to the SHEAR LINE, but also rotate them to allow recession of the sidebar into grooves in the sides of the pins. In locks such as these, one of the mechanisms is known as the primary and the other as the secondary.

probability of detection an assessment, usually numerical on a scale of 0–10, of the likelihood of an intrusion being successfully detected and signalled by a sensor.

processing device an electronic device which is wired between the SENSORS and CONTROL PANEL of an ALARM SYSTEM and which may or may not pass on any signal from the sensors. Signals generated are passed on by the processing device only if pre-set thresholds are met. In an INTELLIGENT ALARM SYSTEM, a processing device may 'look' for a confirmatory signal from another sensor before passing on an alarm signal.

process monitoring the use of an alarm system to monitor activities such as an industrial process or to warn of changes in temperature, electrical overload or problems with boilers or other plant.

processor see PROCESSING DEVICE.

Professional Certification Board see CPP.

profile blocking disc see KEYWAY BLOCKING DISC.

profile cylinder a PIN TUMBLER MECHANISM of uniform cross-section which slides into a LOCK CASE and is secured by a set screw fitted on or under the faceplate or outer FOREND of the lock. A central cam operates the bolt.

P

programmable card an ACCESS CARD which is capable of being programmed or reprogrammed at any time after manufacture.

proprietary (alarm) system a US term to describe an ALARM SYSTEM where the alarm signal is relayed to premises owned and staffed by the company which supplied the alarm system.

proprietary security the supply of security services by a security company under a contract. See also IN-HOUSE SECURITY.

protected area an area that is, or is intended to be covered by an ALARM SYSTEM or enclosed by PERIMETER PROTECTION or is under SURVEILLANCE. Other terms meaning virtually the same include: 'protected location' and 'protected premises'. Some authorities would suggest that an area or building is only 'protected' if it is fitted with an INTRUDER ALARM SYSTEM. The term is also used to describe a location which has been specifically designated (in law) for the purposes of defence or national security. In some countries, if a location is so designated special restrictions apply and, for example, an intruder can be fired upon on sight.

protected location see PROTECTED AREA.

protected premises see PROTECTED AREA.

protection pattern the area which is covered by an individual SENSOR.

protective switch see TAMPER SWITCH.

proximity alarm (system) see CAPACITANCE ALARM.

proximity (card) reader a TAG or CARD READER which is capable of reading the information on certain types of cards at a distance. This development which is based around the TUNED CIRCUIT or SMART CARD is likely to prove a major development in access control technology. See also HANDS-FREE ACCESS CONTROL.

proximity detector (system) see CAPACITANCE ALARM.

PSLC Private Security Liason Council (US).

psychological stress analyser an electronic device with a function rather similar to that of the POLYGRAPH. Those who support the use of this equipment, which is also known as a Voice Stress Analyser, claim that it can detect falsehoods in human speech (either recorded or live) as a result of the presence of certain frequencies and involuntary microtremors. It is claimed that these characteristics are also found in human speech when a lie is spoken. It is interesting to note that most proponents of the polygraph invariably deny that the psychological stress analyser has any significant value in police or security work (and vice-versa).

pull box a type of ALARM STATION where a handle has to be pulled down to activate an alarm. Pull boxes are mainly used in fire alarm systems.

pull button a type of ALARM STATION where a button has to be pulled rather than pushed – this is less likely to be accidentally activated.

pulsed infrared sensor an INFRARED SENSOR which uses pulsed beams for increased coverage and penetration of rain, mist and fog.

punch card a US term for TIME CARD.

punch clock a US term for TIME CLOCK.

punch code the coding on a HOLLERITH CARD, usually by means of holes or indentations.

push button lock a mechanical lock which employs a small dial of buttons on which a combination can be entered. This type of lock is often known as a SIMPLEX LOCK. See also DIGITAL LOCK.

push lock a SUPPLEMENTARY LOCKING DEVICE consisting of a small PIN TUMBLER MECHANISM connected to a hardened steel pin fitted through a brass BOLT. When the CYLINDER is pushed inwards, the lock engages and the bolt is held secure. The device can be fitted to a wide number of types of windows and sliding doors, being particularly suitable for aluminium fittings. Also known as a patio lock or security push lock. (See diagram 8 on p. 68)

PV Positive Vetting, a type of BACKGROUND SCREENING used to provide SECURITY CLEARANCE for very sensitive positions. The 'positive' means that all references, employment history and educational backgrounds are checked to ensure that they are genuine whereas in normal vetting (NV), only negative traces are looked for, such as a criminal record or poor credit history.

pyro-electric infrared sensor a term occasionally used to describe the type of SENSOR also known as a PASSIVE INFRARED sensor.

pyro-electric Vidicon (tube) a closed-circuit television VIDICON with similar characteristics to an IMAGE INTENSIFIER and capable of producing acceptable images in virtual darkness.

Q

quarter-moonlight camera see SIT.

queue a sequence of alarm indications awaiting the attention of an operator in a computer-controlled alarm monitoring system. As alarms are triggered, they will appear on a visual display screen in order, and unless specific priority has been allocated, they will scroll down until they are accepted or cleared by the operator.

quick-response team a special team of security or law enforcement officers trained and equipped to deal with breaches of security. Quick-response teams are required (by the licence conditions issued by the US Nuclear Regulatory Commission) to be available to deal with security incidents at US nuclear power stations.

rack bolt a simple rack-operated device installed in wooden doors as an alternative to conventional surface-fitted BOLTS. Also known as a key-operated mortice security bolt. A similar (but shorter) device is used to secure wooden framed windows and is known as a window bolt. (See diagram 13 below)

rack-mounted a description applied to electronic equipment or components which are of the correct physical dimensions to be installed in standardised racking for easy access and convenient maintenance.

radar sensor a US term (now not generally in use) for a MICROWAVE SENSOR. Also known as a radio frequency motion detector.

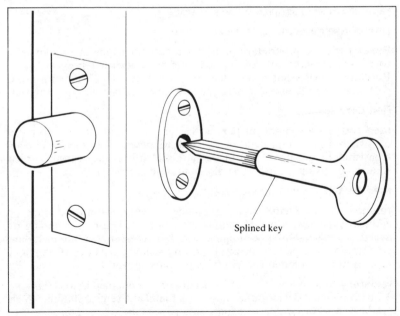

Splined key

Diagram 13 *Rack bolt*

radio alarm system an INTRUDER ALARM SYSTEM in which the individual sensors are not HARDWIRED to the CONTROL PANEL but use radio frequencies instead.

radio frequency analyser see SPECTRUM RECEIVER.

radio frequency motion detector see RADAR SENSOR.

radio telemetry a system using radio frequencies for the collection of alarm data from remote points. The system has many advantages over HARDWIRED data transmission systems, the principal one being cost, particularly when distances are great or no telephone company circuits are available. Although security is generally good, unless the system contains SUPERVISED CIRCUITS there is a danger of JAMMING.

radiowave doppler detector another name for a MICROWAVE SENSOR.

radome the protective housing over a microwave transmitter or receiver.

raid alarm a form of ALARM STATION used to signal a robbery or similar situation. See also PERMANENT CIRCUIT.

raid circuit see PERMANENT CIRCUIT.

random interface a simple scanning technique used on smaller closed-circuit television systems where images are displayed on a MONITOR in a predetermined sequence which is not under the operator's control.

rape alarm see PORTABLE DURESS ALARM (1).

rate-of-rise detector see THERMAL SENSOR.

Razor Tape the proprietary name for a particular type of BARBED TAPE used as a replacement for barbed wire to ENHANCE fences or walls. Recently, it has been noted that the term 'razor tape' is being used, particularly by the press, to describe any improved barbed-type material.

Red Care see ABC.

reed capsule a component part of the most common type of CONTACTS or switch used to protect doors and windows. The reed elements (consisting of two fine metal blades) are sealed in a glass capsule which is dirt and moisture proof. See also MAGNETIC CONTACT.

reed switch see MAGNETIC CONTACT.

refuge a highly secure area resembling a VAULT, within an office or embassy premises, in which staff will be protected from kidnapping or assault. Refuges should be supplied with the necessaries of life including air filtration equipment, back up power, sanitary facilities, water and food as well as communications with the outside world.

register 1. one of a number of books or LOGS maintained in a GATEHOUSE **2**. an obsolescent device occasionally still found in a central station, which produces a record of alarm signals on a paper tape.

Rehabilitation of Offenders Act 1974 a piece of UK legislation which deems certain criminal offences 'spent' after specified periods. It is therefore possible for offenders whose convictions are 'spent' to state without fear of penalty that they have not been convicted of any criminal act. The Act also reinforces the penalties for disclosing the existence of a criminal record to unauthorised persons.

re-key to change the existing combination of a cylinder or lock. Also used in connection with reprogramming CARD ACCESS SYSTEMS.

relocking device a mechanism which DEADLOCKS the bolt mechanism of a SAFE or VAULT door in the event of a physical attack on the lock.

relocking trigger a mechanism integral with a KEYLESS COMBINATION LOCK which DEADLOCKS the lock BOLT in the event of a physical attack.

remote keypad a KEYPAD used to switch on or off a system which is located some distance from the CONTROL PANEL.

R

remote signal a warning or alarm signal which may be transmitted by DIRECT LINE SIGNALLING or other means from the PROTECTED AREA to a CENTRAL STATION or other monitoring point.

remote station an auxiliary CONTROL PANEL, duplicating some or all of the features of a control panel, which is located some distance from the rest of the system. See also REPEATER (2).

remote station system see CENTRAL STATION SYSTEM.

remote terminal an alarm terminal which is usually unmanned and which is directly connected to a CENTRAL STATION.

removable core lock a lock whose CORE can be removed using a CONTROL KEY and quickly replaced thus obviating the need to change the entire lock if a key is lost.

repeater 1 a device similar to a LINE AMPLIFIER
2. a type of REMOTE STATION which displays the signals found on the main CONTROL PANEL. Repeaters are more common in fire alarm technology where they are useful for fire brigade personnel arriving at a site or building with many entrances.

reporting line see ALARM TRANSMISSION CIRCUIT.

reserved keyway lock see RESTRICTED KEYWAY LOCK.

reset to restore an ALARM SYSTEM to its original condition (after an alarm or other actuation).

resolution a measure of the ability of a television system, and in particular the MONITOR, to reproduce detail.

response time the anticipated or actual time in which it is expected that security or police personnel will arrive at a location following the triggering of an alarm.

restricted keyway (lock) a lock, usually made to special order, where the KEYWAY is designed to take a blank or special cross-section. This restricts the copying of keys, which are registered and normally supplied only to the original customer. This type of lock is also known as a reserved keyway lock.

retail security the protection of assets and prevention of theft within retail premises, with particular reference to the prevention and detection of SHOP THEFT and theft by employees.

retractable door cord a specially made door cord which is designed to fit overhead or sliding doors.

RGS Remote Ground Sensor: a portable SEISMIC DETECTION SYSTEM, originally developed for the military. The system is unique in that each sensor is combined with a VHF transmitter linking the sensor to the control equipment. The equipment is easily man-portable in one or two satchels. See also CLASSIC.

rhodamine a moisture-activated VISIBLE STAINING COMPOUND producing a bright red colouration on skin contact.

rim automatic deadlatch see RIM AUTOMATIC DEADLOCK.

rim automatic deadlock a DEADLOCK offering a higher degree of security than other RIM LOCKS. This type of lock will automatically secure itself when the door closes. (See diagram 12 on p. 71)

rim cylinder lock see RIM NIGHTLATCH.

rim lock a lock made for fixing to the inner surface of a door.

rim nightlatch a NIGHTLATCH made for fixing to the surface of a door; its STAPLE or STRIKING PLATE is also surface-mounted on the door frame. (See diagram 11 on p. 71)

ringback an acknowledgment (usually audible) from a CENTRAL STATION to a SUBSCRIBER that a closing or OPENING SIGNAL has been received.

ringing the changes a simple fraud in which the perpetrator (usually with an accomplice) confuses or distracts a shopkeeper or trader so that excessive amounts of change are obtained.

rising step barrier see HYDRAULIC VEHICLE BARRIER.

risk management the application of insurance and other techniques to determine the extent of possible loss facing a company. A risk management study would normally include a detailed evaluation not only of security precautions but also of fire and safety measures.

RL Registered Locksmith, the first level of proficiency in a professional registration program organised by the ALOA.

road blocker see HYDRAULIC VEHICLE BARRIER.

rolled glass see ANNEALED GLASS.

roller a hardened steel bearing-like component found set into the DEAD-BOLT of good quality MORTICE LOCKS, which prevents sawing through the bolt. Also known as an anti-cutting roller. (See diagram 9 on p. 69)

roller bolt a SPRING BOLT with a ROLLER on a pin rather than a bevel. Used to facilitate installation of locks on the right or left of doors. A roller bolt is also less likely to injure children or damage clothing. Often used in cupboards.

roller shutter a strong shutter, usually metal and constructed from inter-linked slats, often used to secure the exterior of shops. Roller shutters offer excellent security, particularly for large expanses of glass. Decorative shutters of wood or light metal are often used in hotels, bars and shopping arcades to restrict access when parts of the premises are closed.

R

rose an ornamental plate or ring on the outer surface of the door, often around the handle. (See diagram 9 on p. 69)

rotating disc lock a lock whose CYLINDER contains a set of flat disc DETAINERS which must be rotated by a key to the correct position for a SIDEBAR to enter notches or grooves on the discs.

rough cast glass see ANNEALED GLASS.

round see BOLT.

round key see TUBULAR KEY.

S

sabotage the deliberate and planned destruction of property or plant, normally with the intention of disrupting the activities of a company or the functioning of a service.

Sabre Tape the proprietary name for the FIBRE OPTIC TAPE used for FENCE protection, originally developed by Pilkington Security.

safe a secure container for money and valuables. Safes may be free-standing, bolted to the floor, bricked-in, fitted within walls or sunk into floors. The best safes will offer a degree of fire-resistance to their contents and should be heavy enough to resist removal (to a quiet place for leisurely opening). There are insurance industry standards for safes in the US (see Appendix 2), but unfortunately the UK safe manufacturers have so far resisted any form of consensus on nationally-accepted standards.

safe alarm a device consisting of a CONTACT MICROPHONE or CAPACITANCE SENSOR which can be fitted to a safe combination dial or in front of the KEYWAY. Also sometimes known as a SAFE COVER or limpet. (See diagram 2 on p. 29)

safe cover a magnetic device, sometimes known as a limpet, which operates on the CAPACITANCE principle. The device is placed over the combination dial or KEYWAY of a safe. If the cover is touched without first switching off the circuitry an alarm is triggered.

safe deposit lock a lock used in safe deposit boxes and lockers of the type usually installed in a strongroom or VAULT. Such locks have two key mechanisms; a bank official or guard must insert and turn his key (sometimes called a guardian key) first. This allows the subsequent operation of the renter's key. One guardian key usually operates all the boxes, each of which has an individual CHANGE KEY for the renter. In the most modern facilities, the guardian key function is performed electrically from a central reception desk as soon as a renter has identified himself.

sash lock 1. an upright MORTICE LOCK, usually with a narrow cross-section so as to fit thinner door jambs or window frames. This type of lock is also known as a narrow stile lock
2. the term is also used (incorrectly, according to some authorities) to refer to the SUPPLEMENTARY LOCKING DEVICE better known as a PUSH LOCK, when that device is fitted to sash windows.

102

sash ward a type of WARD or obstruction around the keyhole used in MORTICE or RIM LOCKS either alone or together with LEVERS to obtain an increase in the number of DIFFERS.

sash ward cuts the CUTS made in the BLADE of a LEVER LOCK KEY to bypass the SASH WARDS. (See diagram 6 on p. 61)

sash window a wooden-framed window, usually in two sections, which moves vertically. When in the closed position, sash windows are usually locked with a FITCH CATCH that offers poor security. This type of catch should be replaced by a LOCKING CAM CATCH or supplemented with a DUAL SCREW, or the window should be permanently secured through the frames.

satellite see COLLECTOR.

satellite camera a concealed closed-circuit television camera housing, which is hemispherical and fitted with a number of dummy lenses and flashing red lights. The unit is fitted to the ceiling of a location. It is impossible for a casual observer to determine whether or not a real camera is installed in the satellite housing and it is therefore possible to move a few real cameras around a larger number of satellite housings.

scanning receiver see ANTI-EAVESDROPPING DEVICE, SECURITY SWEEP.

scan period the total amount of time the monitoring circuitry of an electronic system takes to interrogate or poll each ZONE or circuit to determine whether any change in status has taken place since the last scan took place. Also known as the polling period.

scene analyser a type of MOTION SENSOR which not only draws the attention of the operator to a changing pattern on a MONITOR but also marks the disturbances on the screen – usually with a flashing pattern.

scissor-action stay a type of STAY made of pivoting metal sections, which is used to hold a top-hung CASEMENT window in the open position.

scrambler a popular term for any telephone fitted with a speech distortion device.

search register one of the LOGS maintained in a GATEHOUSE. It should contain details of all searches of individuals or vehicles performed at that location.

secondary locking mechanism see PRIMARY LOCKING MECHANISM.

secure mode the condition of an ALARM SYSTEM in which all the sensors and circuits are operational and which is not triggered.

secure telephone system a telephone system which incorporates signal encryption and/or line supervision to prevent interception of conversation.

securing device a term used to encompass all types of locks, catches, bolts and similar devices.

103

security bar a metal bar used to secure an access door, hatchway or filing cabinet, usually fixed at one end with a PADLOCK. Security bars are very commonly used, as an alternative to chains, to secure fire exits and PANIC BARS when premises are empty.

security classification a system of classifying documents according to their contents so that a piece of information (such as the formula of a new drug) on which a company's future might depend would be classified 'secret' whereas the new price list might be only 'company confidential'.

security clearance a type of BACKGROUND SCREENING. The term is normally used in the police or military sense, where such screening is done with a view to obtaining permission for an individual to work on a defence-related project.

security company a company supplying security-related products or services.

security fence see FENCE (1).

security glazing the term is used, somewhat loosely, to describe different types of glazing material (mainly laminates of glass and polycarbonate) that are used to prevent the passage of missiles or attacks on staff. See also BANDIT-RESISTANT GLAZING, BULLET-RESISTANT GLAZING.

security industry a term normally applied to include all those companies involved with security equipment or supplying security services as well as agencies supplying investigative and similar services.

security lighting an essential part of any well thought-out security plan. It is vital to ensure that specialist advice is sought to obtain the best balance of light levels, lamp life, colour rendering and running costs. Where a site is to be fitted with closed-circuit television, this must be taken into account when planning the lighting. See also GLS, LUX, MBF, MBI, MCF, SON, SOX, TH.

security line see ALARM TRANSMISSION CIRCUIT.

security loop see ALARM TRANSMISSION CIRCUIT.

security monitor a device which supervises the communications circuitry and line of an ALARM SYSTEM. See SUPERVISED CIRCUIT.

security sweep an organised search for illicit listening devices or BUGS, using a range of electronic equipment. See also ANTI-EAVESDROPPING DEVICE, CARRIER-CURRENT RECEIVER, NON-LINEAR JUNCTION DETECTOR, SPECTRUM RECEIVER, TELEPHONE ANALYSER, TIME DOMAIN REFLECTOMETER.

security system any combination of electronic, electrical or mechanical equipment used to protect persons or property. The term can also be applied to plans and procedures used for the same purpose.

SEIA Security Equipment Industry Association (US).

seismic detection (system) a PERIMETER DETECTION SYSTEM using buried devices such as AUDIO DETECTION or GEOPHONES.

self-contained card reader a CARD READER which is constructed with its own internal microprocessor, power supply and associated circuitry. Such units do not require connections to other control equipment. Reprogramming is normally done by a hand-held plug-in KEYPAD.

self-contained intruder alarm (system) a unit designed for domestic use which consists of (usually) a PASSIVE INFRARED SENSOR together with a SOUNDER, a power supply and a KEY SWITCH. These units are often enclosed in a veneered box and resemble small stereo system loudspeakers. When SET, any activity in the room which contains the system will result in the alarm being triggered and the sounder activated – with the aim of scaring the intruder into fleeing. Another piece of similar equipment, sometimes called a Notecalm alarm (after the principal UK manufacturer) is fitted with powerful flashing lights as well as a sounder.

self-destructing (paper) badge see BADGE.

self-locking latch see AUTOMATIC DEADLATCH.

sensing wire see E-FIELD SYSTEM, FIELD WIRE.

S

sensor a device which responds to a specific stimulus or event and then passes on a notification of that event to a CONTROL PANEL. See also DETECTOR.

sequential card reader a CARD READER which contains a KEYPAD. Such units require not only the insertion of a card but verification of the cardholder's identity by entering a PIN.

sequential switcher a device used to display closed-circuit television images from more than one camera one after another on a selected MONITOR. The sequence of images and the duration for which the image is displayed (the dwell time) can normally be adjusted to suit the user.

set a term used to describe an INTRUDER ALARM SYSTEM which is switched on and ready to respond to any attempt to enter the PROTECTED AREA. A system which is switched off is said to be 'unset'.

SFPE Society of Fire Protection Engineers (US).

shackle the hinged, sliding or swivelling part of a PADLOCK which loops through a fastening device and which is then secured within the padlock body.

shadow see STRAWMAN.

shank the body or shaft of a LEVER LOCK KEY; strictly, that part between the BOW and the COLLAR. (See diagram 6 on p. 61)

shatterproof a term with no precise definition, often found in advertising material for glass. It is normally taken to mean that if the glass is broken

105

it will not splinter or break into sharp pieces. If glass or other material is to be rendered shatterproof, a thin coating of plastic or polythene-like material is laminated to the inner surface to prevent fragmentation after breakage of the glass. See also BANDIT-RESISTANT GLAZING, LAMINATED GLASS, TOUGHENED GLASS.

shatter-resistant film a type of polyester film, usually 0.05–1 mm thick, used to increase the resistance of 'normal' glass to impact. Such films also tint or darken glass and assist in keeping out solar radiation.

shear line the line of the circumference of the PLUG in the SHELL of a PIN TUMBLER CYLINDER. This is the location at which the PINS (or other TUMBLERS) must be aligned so as to inhibit their obstruction to the rotation of the plug and hence the opening of the lock. (See diagram 10 on p. 70)

sheet glass see ANNEALED GLASS.

shell the external CASE of a lock CYLINDER, that is, less the CORE or PLUG. (See diagram 10 on p. 70)

shimming see LOIDING.

shock sensor see PIEZOELECTRIC DETECTOR, VIBRATION DETECTOR.

shoot see BOLT THROW, LOCK SHOOT.

shoplifting a euphemism for theft from retail premises. The term is normally used to refer to theft during opening hours by person entering the premises as customers.

shop theft a term used to cover all theft from retail premises including acts by staff.

shot the closed or secure position of any BOLT or LOCK.

shoulder the ear or projection on a CYLINDER KEY which prevents it being inserted too far into the lock. The COLLAR of the LEVER KEY performs a similar function. (See diagram 5 on p. 61)

shredder a mechanical device used to render documents and other materials (typewriter ribbons, microfiches etc.) unreadable by cutting the material to be shredded into narrow strips. Conventional office shredders do not achieve an adequate level of illegibility for security purposes and in this case, higher security models should be used which cross-cut the strips into fragments. Alternatively, consideration should be given to other means such as pulping or pulverisation.

shrinkage a euphemism used in retail security to describe the losses of stock due to employee and customer theft, breakages and other waste.

shunt to bypass an alarm or alarm ZONE by a special circuit or switch. See also ALARM BYPASS.

shunt lock a specially modified lock which normally acts as a FINAL EXIT lock and which is connected to the alarm system to SHUNT out that ZONE from the system. Also known as a LOCK SWITCH.

shunt switch a KEY SWITCH (or other secure switch) used instead of a SHUNT LOCK.

shut-out key in hotel keying systems this is usually the EMERGENCY MASTER KEY which can make a lock inoperative to all other keys. The shut-out key is used for additional security on special occasions, to preserve a crime scene or to impound a guest's property (until a bill has been paid).

sidebar (lock) a lock in which the CYLINDER contains a length of solid metal which extends over the PINS, DISCS or TUMBLERS and projects outside the PLUG into the lock SHELL or body. As the tumblers line up, the sidebar enters notches or grooves in them to allow the plug to rotate. It may be used as a primary or secondary locking system.

side hook see HOOK AND EYE.

signalling system the component part of an intruder detection device which passes an alarm or other signal from the sensing element to the CONTROL PANEL or which relays a signal from the control panel to a remote monitoring location.

signature verification system a high-security identification system which digitises and stores the dynamic characteristics of a handwritten signature (or other pre-programmed phrase). This is one of the fairly recent BIOMETRIC SENSOR systems which should offer useful security applications, particularly in the banking field.

silent alarm an alarm without a local sounder. Sometimes also used to describe any alarm where the local sounder is programmed not to ring for a predetermined period to allow police a chance to respond before an intruder is alerted.

silicon intensified tube see SIT.

simplex the transmission of a single alarm or other signal down one circuit. See also DUPLEX, MULTIPLEX.

Simplex lock the proprietary name for a popular and efficient mechanically-operated DIGITAL LOCK often used in medium security applications. The term is often used to refer to any lock of this type.

simulated camera a low-cost, dummy closed-circuit television camera used alone or in conjunction with real cameras to give the impression that surveillance coverage exists or is more comprehensive than it really is.

single-entry system another term for a system fitted with an ANTI-PASSBACK CIRCUIT.

single-voiding card reader a STAND-ALONE CARD READER which is capable of cancelling individual ACCESS CODES thus barring them from the system.

S

107

SIT Silicon Intensified Tube: the essential component of a type of LOW-LIGHT TELEVISION CAMERA whose ability to produce acceptable images at night and at light levels from 2–0.01 LUX derives from a component known as a cascaded intensifier. An ISIT is an augmented version of the SIT, and is fitted with two or three intensifier stages rather than the single stage fitted in the SIT, permitting it to operate at 0.001 lux. SIT cameras are often called 'quarter moonlight' units while ISIT cameras are called 'starlight' units. See also NEWVICON INTENSIFIED TUBE.

skeleton key a key with a BIT or BLADE that has been deliberately cut away to allow it to bypass WARDS or obstructions within a lock.

skipper an individual who evades payment of a hotel or restaurant bill by leaving surreptitiously. Often used to describe a criminal who habitually obtains free lodging in this way.

SLA Security Lock Association (UK).

slam action lock any lock which automatically secures itself when a door is closed.

sleeping policeman see SPEED BUMP.

SLI Security Law Institute (US).

slowscan (closed-circuit television system) a technique for transmitting closed-circuit television signals down telephone wires. The image is digitised, transmitted and then reassembled at a remote point. At the present time, circuitry limitations mean that the image is still rather than moving, and some seconds old.

smart card an ACCESS or other ENCODED CARD which contains its own circuitry and memory.

SMNA Safe Manufacturers' National Association (US).

snap lock a SUPPLEMENTARY LOCKING DEVICE consisting of a small PIN TUMBLER MECHANISM connected to a self-locking claw device. The unit is used to secure wooden CASEMENT windows. The key is used only to unlock the device.

snib the catch on a RIM NIGHTLATCH or RIM AUTOMATIC DEADLOCK which locks the BOLT in either the locked or withdrawn (open) position. (See diagrams 11 and 12 on p. 71)

sniffer 1. a device used to search baggage or freight by drawing in samples of air and analysing them for the presence of explosive vapours
2. a device used during alarm surveys, capable of locating the presence of existing ultrasonic frequencies which might cause false alarms when an ULTRASONIC MOTION DETECTOR system is installed.

software the programs or microchip-encoded information which control the functioning of electronic or microprocessor-based systems.

SON a high-pressure sodium lamp which is economical to run, producing a relatively high amount of light per watt and having a long life.

sonic motion detector a comparatively rarely-used SENSOR which uses audible sound waves to detect the presence of an intruder using disturbances in the sound wave pattern and reflections in the DOPPLER SHIFT.

sound discriminator a type of alarm SENSOR which responds to specific sound frequencies, for example a GLASS BREAK DETECTOR.

sounder a noise-making device. The term encompasses banshees, bells, hooters, klaxons, sirens, horn speakers and YODELARMS.

sound sensor an alarm SENSOR which uses a microphone to detect sounds and then triggers an alarm if those sounds fall within a pre-set range.

SOX a low-pressure sodium lamp which produces the most light for a given power input.

space protection the use of a variety of different types of SENSORS (mainly VOLUMETRIC SENSORS) to protect a large open area, for example a warehouse. Also known as area protection. (See diagram 2 on p. 29)

span (sensor) a type of SENSOR which provides detection cover within a straight and narrow field. Examples of this type of sensor would include all PHOTOELECTRIC SENSORS.

spectrum receiver an electronic device used in the detection of illicit listening devices which monitors frequencies from the very low to the very high range of the radio spectrum (10 KHz–2 GHz). Also known as a radio frequency analyser. See also SECURITY SWEEP.

speed bump a low obstruction on a roadway, intended to slow traffic. Speed bumps may be constructed of tarmac or may be prefabricated from rubber or other durable material. Also known as a sleeping policeman.

SPI Society of Professional Investigators (US).

spindle a metal bar, or square cross section which passes through the FOLLOWER of a LATCH to allow the handle or knob to operate the latch bolt. (See diagram 9 on p. 69)

spindle hole the hole in the FOLLOWER through which the SPINDLE fits. (See diagram 9 on p. 69)

splined key a key-like device used to secure some types of SUPPLEMENTARY LOCKING DEVICE. (See diagram 13 on p. 97)

split-image lens a special type of lens which allows the viewing of two different fields or television images (usually on the same MONITOR).

split screen a technique whereby more than one closed-circuit television camera's output can be displayed on a single MONITOR's screen. Current

S

equipment is capable of displaying up to four separate images or scenes on a single monitor.

spool pin a top or DRIVER PIN in a PIN TUMBLER LOCK, which is shaped like a spool to give the lock increased PICK resistance. See also HUCK PIN, MUSHROOM.

spot protection the protection of individual objects by dedicated SENSORS (for example, paintings and safes). The phrase is also sometimes used to describe the protection of a confined space. Also known as point or object protection. (See diagram 2 on p. 29)

springbolt a BOLT which is spring-loaded and therefore, always LIVE. Another name for a LATCH BOLT.

spring contact a switch-type SENSOR which used to be frequently employed to protect doors and windows. It works in the same way as the plunger switch used on car doors to switch on the courtesy light and is easily countered. A similar sort of device can be used for protecting paintings.

spring cover the plate which secures the springs and PINS in a PIN TUMBLER MECHANISM. (See diagram 10 on p. 70)

spring latch see SPRING LOCK.

spring-loaded glass bolt see GLASS BOLT.

spring lock a lock which is usually fitted with a bevelled LATCH BOLT which allows the door to be opened from one side with a knob, but only by means of a key on the outside. Frequently used on, for example, fire exits.

spring-shackle padlock a PADLOCK which is fitted with a spring mechanism which causes the shackle to spring open when the key is turned. This type of padlock also has a SPRINGBOLT so the shackle locks automatically when pushed home.

spyhole see DOOR VIEWER.

SSTV Slow Scan Television see SLOWSCAN.

stand-alone (system) a piece of equipment which is self-contained and not connected to any other larger system.

stand-by battery see BACK-UP BATTERY.

staple 1. a box-like fastening surface mounted on a door jamb into which the bolt of a RIM LOCK shoots. See also STRIKING PLATE. (See diagram 11 on p. 71)
2. see HASP AND STAPLE.

starlight (camera) a term used to describe the ultra low-light capabilities of the ISIT television camera. See SIT.

starlightscope see IMAGE INTENSIFIER.

static post see MANNED POST.

stay see WINDOW STAY.

stay bolt a SUPPLEMENTARY LOCKING DEVICE used to secure a WINDOW STAY in the closed position. It consists of a PIN TUMBLER-operated bolt, which when it is SHOT secures the stay arm in the closed position.

stay lock a SUPPLEMENTARY LOCKING DEVICE used to secure a CASEMENT window, consisting of a threaded section (which replaced the original STAY PIN) and a locking nut which is operated by a simple key. Also known as a casement stay screw. (See diagram 1 on p. 19)

stay pin see LOCKING STAY PIN.

stile the vertical edge and face of a door, particularly the part where a MORTICE LOCK is installed.

strain gauge sensor see PRESSURE SENSOR.

strawman a strawman is a non-existent employee, sometimes known as a ghost worker or shadow, whose name appears on a payroll and whose salary is fraudulently claimed and collected by a site manager or other person in a similar position of trust.

stress sensitive cable see TAUT WIRE SYSTEM.

stress sensor a SENSOR usually used in TAUT WIRE SYSTEMS, which responds to an increase or decrease in pre-set loadings.

strike see STRIKING PLATE.

strike plate see STRIKING PLATE.

striking plate a flat metal plate fixed to the door jamb which the latchbolt of a lock 'strikes' as the door shuts. It will contain one or more bolt holes (a DEADBOLT hole and a LATCH BOLT hole). If these are boxed-in to protect the bolt from end pressure then the device could also be called a box striking plate. (See diagram 9 and 11 on p. 69 and p. 71)

strobe light a light possessing a very bright flash, often used as a visual accompaniment to an audible alarm.

strongroom see VAULT.

stump see BOLT STUMP, PIVOT, POST.

sub-master (key) see MASTER KEY.

subscriber an individual or organisation using the services of a security company. The term is usually used to denote the rental of an alarm system but the word is also used to describe any person or company paying a maintenance, CENTRAL STATION or KEY HOLDING fee.

subscriber unit the term used by UL to describe an AUTHORISED ACCESS CONTROL SWITCH.

supervised circuit any circuit that is capable of detecting and signalling abnormal line conditions. These conditions may include power loss, short circuit or even JAMMING. The best systems will indicate a circuit fault at both the protected premises and the CENTRAL STATION. If the central station receives such a signal, they will normally attempt telephone contact and if this is not possible, they will either notify the police or ask a security patrol to investigate.

supervised closing see SUPERVISED OPENING/CLOSING.

supervised entrance an entrance to a building which is continuously monitored either physically by an attendant or remotely, using closed-circuit television.

supervised line see SUPERVISED CIRCUIT.

S

supervised opening/closing a method of alerting a CENTRAL STATION to activity within premises fitted with an alarm. If the alarm is activated outside a very narrow range of agreed times, a police or security response is dispatched. See also OPENING SIGNAL.

supervision a term used to describe the circuits or lines connecting an ALARM SYSTEM or its CONTROL PANEL to a CENTRAL STATION (or other location), when those circuits are continuously monitored. See also SUPERVISED CIRCUIT.

supplementary locking device an additional lock or catch fitted to an existing securing device on a door or window. This term is often used in respect of window catches installed to supplement the non-security closing devices supplied by the window manufacturer or installer.

surveillance the process of controlled observation, either covert or highly visible. Surveillance may be by human eye, by CAMERA or by some electronic SENSOR. Surveillance also encompasses the monitoring of physical conditions such as temperature, pressure and the correct operations of plant and machinery.

surveillance camera any CAMERA that is used to watch either constantly or intermittently over a PROTECTED AREA.

suspicion alarm a device similar to a HOLDUP ALARM, which can be covertly activated by a cashier to alert a remote location or other staff to a potential problem. Suspicion alarms can also trigger ARMOURED SHUTTERS or BANK CAMERAS.

suspicion button see SUSPICION ALARM.

suspicion camera see BANK CAMERA.

sweep a term used to describe the process of detecting a surveillance or

EAVESDROPPING DEVICE – a more 'acceptable' term than DEBUG. See also
SECURITY SWEEP.

swipe reader see WIPE-THROUGH CARD READER.

synoptic map see MIMIC PANEL.

S

T

tag a passive (non-powered) device used in EAS systems, sometimes referred to as a token. The tag operates by triggering an alarm when a garment (or other item) to which it is attached is brought into the range of a sensing signal emitted between two free-standing pillars, or between sensors built into a doorway.

tagging the act of securing or identifying equipment or stock by **1.** fixing a TAG to it
2. marking equipment with a CHEMICAL TAGGING system.

tailgating see PIGGYBACK.

tailpiece a (usually) flat piece of metal fixed to the rear of the PLUG which transmits the turning motion of the plug to the BOLT mechanism in the lock CASE. Sometimes referred to as the connecting bar.

tally clerk see CHECKER.

tamper an attempt to interfere with or bypass a security system or any of its components.

tamper switch a type of switch (usually a mechanical microswitch) designed to trigger an alarm if an unauthorised attempt is made to remove the cover or gain access to any component part of a security system.

tap 1. to covertly attach a listening device to a telephone instrument or line
2. the device itself. See also TELEPHONE TAP.

tape dialler a communications device which was in considerable use before the development of the DIGITAL DIALLER. It consists of a telephone pulse or tone dialler and a tape player. If the alarm system to which the tape dialler is connected is triggered, the unit will dial either the police emergency number (or any other preprogrammed number) and then play a prerecorded message stating the fact that intruders are at the specified location. These systems have only a low level of security as they can be defeated by dialling-in to the telephone line to which they are attached. Before the widespread use of tape as a recording medium, small gramophone or phonograph records were utilised to carry the alarm message.

target hardening a military phrase adopted by security and used to refer

114

to the improvement of physical security in a given location. It usually refers to ENHANCEMENT of security against a given threat.

target integration a system for increasing the sensitivity of a closed-circuit television camera where the camera is used for continuous observation of a motionless scene. The camera's VIDICON is cut off for a predetermined period and when the image is picked up immediately after it is turned on again, it is scanned line by line to increase the RESOLUTION of the image.

target-to-background differential a term used with relation to the selection of PASSIVE INFRARED sensors. The term refers to the difference between the AMBIENT TEMPERATURE and that of an object or target within the PROTECTED AREA.

taut wire system a PERIMETER PROTECTION SYSTEM using mechanical components. When a fence is protected with this system, it is fitted with a predetermined arrangement of wires, springs and strain gauges. Any attempt to cross, climb or cut through the fence increases or decreases tension on the system triggering an alarm.

TDR Torch- and Drill-Resistant: a term used to indicate that a safe has been constructed from TDR metal alloys.

telecontact a procedure whereby a guard or watchman is required to phone (or respond to a call from) a designated number at regular intervals during a shift. This system, often used in MUTUAL AID schemes is also provided for a fee by most CONTRACT SECURITY companies or commercially operated CENTRAL STATIONS. Failure to call-in or to respond to a call will actuate a predetermined response either from police or a security supervisor.

telemetry the collection and transmission of data from remote units to a central processing system.

telephone analyser a device capable of detecting if electronic eavesdropping equipment is connected to any part of a telephone circuit.

telephone dialler see TAPE DIALLER.

telephone entry system see DOOR PHONE.

telephone line monitor a device which initiates an alarm if a telephone circuit to which it is attached is TAMPERED with. See also SUPERVISED CIRCUIT.

telephone scrambler a device used to make a telephone conversation incomprehensible to any other party listening to the call. Simple, commercially available scramblers work on the principle of speech frequency inversion and are of little security value.

telephone tap a device used to intercept any communication between telephones. A tap may be within the telephone instrument (when it is usually referred to as a BUG) or elsewhere in the circuit, when it is referred to as an IN-LINE PHONE TAP. The act of fitting the device is known as 'tapping'.

tempered glass see TOUGHENED GLASS.

terrain-following sensor a SENSOR which is capable of adjusting the area protected to the natural contours of the ground using aerials or reflecting devices – normally an INFRARED or MICROWAVE SENSOR.

test position see KEY SWITCH.

test purchase a technique for verifying compliance with retail procedures. Security or audit personnel not known to shop staff will visit a retail outlet and make purchases. Later, receipts are checked against AUDIT ROLLS or other sales records to ensure that the cashier has rung up the correct amount, given the correct change and followed all laid down procedures. The technique is also used to ascertain the attitude of staff to customers. See also DISHONESTY TESTING.

TH a tungsten halogen lamp used extensively for short-range floodlighting, particularly around gates and GATEHOUSES. It has a high level of output but also a high level of energy consumption.

theft of service see ILLEGAL ABSTRACTION.

Theory of Interchange see LOCARD'S PRINCIPLE.

Theory of Transference see LOCARD'S PRINCIPLE.

thermal burning bar this is an improved, updated form of THERMIC LANCE where the mild steel rods in the pipe are replaced by rods containing a combination of magnesium and other alloys which burn at even higher temperatures.

thermal imager a HAND-HELD device using THERMAL IMAGING circuitry to provide vision inside smoke-logged buildings and for rescue purposes, when the device enables firemen to pinpoint the presence of casualties trapped in damaged or collapsed buildings.

thermal imaging a technique based on the use of complex scanning circuitry and germanium lenses which enables the equipment to amplify heat emitted in the form of thermal electromagnetic radiation by various objects. The pattern formed is 'scanned' and its INFRARED intensity is recorded and converted to a brightness value. This is then displayed on a monitor or screen as a 'picture' of the scene.

A portable THERMAL IMAGER is currently in production; although it is still comparatively bulky, this is likely to change and lightweight units, possibly capable of being fixed to a helmet or worn like infrared goggles, will soon be a commercial proposition.

thermal sensor a SENSOR which is designed to respond when a specified AMBIENT TEMPERATURE is recorded. Thermal sensors can also be programmed to respond when a specified rate of temperature increase is noted over a pre-set period; these are known as rate-of-rise detectors.

thermic lance a type of industrial cutting tool which was successfully used to attack many older safes and vaults in the 1970s. It consists of a long steel pipe packed with mild steel rods connected to an oxygen supply. When the end of the pipe is heated red hot, the oxygen is turned on causing the whole device to burn rapidly at a very high temperature. Normally used on construction and demolition sites, the use of the thermic lance is restricted in confined spaces both by the weight of the equipment and the high volume of toxic fumes produced by the process. See also THERMAL BURNING BAR.

thermistor circuit a component circuit used in ULTRASONIC MOTION DETECTORS to prevent a DETECTION RANGE shift due to AMBIENT TEMPERATURE changes.

throat the part of a LEVER LOCK KEY between the COLLAR and the BIT. (See diagram 6 on p. 61)

throughput rate the number of individuals per hour capable of being processed by an ACCESS CONTROL SYSTEM.

throw see BOLT THROW, LOCK SHOOT.

thumb knob see SNIB.

thumb turn a fitting on the inside of a lock used to operate the bolt (often found in, for example, lavatories).

tilt switch a type of ELECTROMECHANICAL DETECTION DEVICE where the movement of a blob of mercury in a sealed glass tube will make a connection across two conductors if the tube is tilted. An INERTIA SWITCH will usually fulfil the same role (except in the case of vehicle alarms where inertia switches are designed not to respond to tilting resulting from high winds or parking on a slope).

time card a card used in a TIMECLOCK.

timeclock a device used to record the arrival at and departure from work of (usually) hourly paid employees. Each employee inserts a card into the timeclock and the appropriate time is stamped on the card. This card is then used to calculate wages. See also CLOCKING ON, CLOCKING FRAUD.

time/date generator an electronic device which is used to imprint the time and date on to a video recording for later identification. Similar devices are used in some film cameras for dating and identifying photographic prints or slides.

time delay circuit see DELAY CIRCUIT.

time division multiplexing a method of MULTIPLEX SIGNALLING using the natural gaps between signals.

117

time domain reflectometer an electronic device (used in the search for illicit listening devices) which detects telephone line faults. Such faults may indicate the presence of an IN-LINE PHONE TAP. Also known as an echo pulse fault locator.

time keeper an employee whose job includes the supervision of employees CLOCKING ON and off and whose duties may also include the surveillance of the employee entrance to a building.

time lapse a circuit fitted to any type of image-recording device which permits intermittent but regular recording on film or video tape.

time lock a device fitted to vaults, strongrooms and larger safes in addition to normal locks. Time locks provide additional security in that they permit the door to which they are fitted to be opened only within a pre-set time frame, even if the correct key or combination are available. This prevents any attempt to use an illicitly obtained key and minimises the possibilities of gaining access by duress outside normal working hours.

time zone the times between which an individual is permitted to a location entry via an ACCESS CONTROL SYSTEM. Time zones will normally be allocated on ENROLMENT. See also ACCESS CARD.

tip the end of a CYLINDER LOCK key which engages and turns the CAM. (See diagram 5 on p. 61)

Title 31 see BANK SECRECY ACT.

T-line the proprietary name for a PERIMETER PROTECTION SYSTEM which resembles the E-FIELD SYSTEM.

TOBIAS a proprietary GEOPHONE system originally produced to meet a military need. It is fully portable but unlike the RGS system it requires cable to be laid from each sensor to the control equipment.

toe switch see FOOT SWITCH.

token a term used to describe one type of access device used in a HANDS-FREE ACCESS CONTROL SYSTEM. Tokens differ from TAGS in that they are normally self-powered, drawing supply from an internal, long-life mercury cell or from a rechargeable battery.

top pin the driver or uppermost PIN touching the spring in a PIN TUMBLER CYLINDER. (See diagram 9 on p. 69)

toughened glass a type of specially treated glass (also known as tempered glass) which has a greater resistance to impact than ANNEALED GLASS.

tour clock see WATCHMAN'S CLOCK.

tower bolt a surface-mounted, non-locking securing device in which the BOLT is held in place by two or more straps (as opposed to the usual barrel).

transceiver a transmitter and receiver housed in a single unit.

transom a lintel or member running horizontally across a window.

transom lock a device using a screw fitting which will secure a TRANSOM WINDOW by clamping the WINDOW STAY to its locating bracket.

transom window a small window above the lintel of a door or above another window. Sometimes used as a synonym for FANLIGHT.

transponder a device which collects SENSOR or other data and converts it into a form suitable for onward transmission.

transverse wiring a CLOSED-CIRCUIT device used in a similar fashion to TUBE AND WIRE where a fine mesh or pattern of thin PVC-insulated wire is installed inside a door or hollow wall. Any attempt to break through the obstacle results in a broken wire which triggers an alarm. Transverse wiring is economical and is often incorporated in the walls or floors of vaults as a back-up to other sensors.

trap protection the use of varying types of SENSOR to provide a second level of detection around a possible target. For example, if a building contains a room in which there is a very high-value object, trap protection could be provided in the corridor outside the room to detect the presence of an intruder who had avoided either the PERIMETER PROTECTION or the door or window sensors. (See diagram 2 on p. 29)

trigger see ANTI-SHIM DEVICE.

trouble signal a warning indication from a SENSOR forming part of an INTELLIGENT ALARM SYSTEM that the sensor is either defective or in need of adjustment. Trouble signals either generate a FAULT indication so that the operator will call for a service technician, or in even more sophisticated systems they can actually generate a specific request for service via the CENTRAL STATION.

tube and batten see TUBE AND WIRE.

tube and wire a simple method of protecting openings such as skylights from intrusion. It uses thin, uninsulated wire run in small diameter metal pipes fixed across the area to be protected. Any intrusion will cause the pipes to bend or break and create a short-circuit triggering an alarm. Tube and wire can also be used to provide protection for door panels or above false ceilings. As this technique invariably uses wooden battens on which the tubes are mounted, it is sometimes known as tube and batten. Also known as lacing.

tubular key a key with a tubular BLADE or cylindrical cross-section. This hollow blade has its CUTS at the end of the blade.

tubular key lock a lock which operates on the PIN TUMBLER MECHANISM principle in which the pins are arranged in a circle facing the front of the

CYLINDER. When the correct key is inserted, it pushes inward to align the pins.

tumbler a tumbler was originally a DETAINER with a stump-like projection which had to be raised before a bolt could be moved. The term has now largely lost this meaning and is now commonly used as a collective term for PINS, LEVERS, DETAINERS, DISC TUMBLERS, wafers and WHEELS as well as other movable obstructions which prevent an incorrect key from operating a lock.

tuned circuit a sensing circuit used in a HANDS-FREE ACCESS CONTROL SYSTEM to read TAGS, TOKENS and SMART or ACTIVE CARDS.

turnstile a mechanical device used to regulate the entry and exit of individuals into a given area. Turnstiles may be half-height, like those at the entrances to sports grounds or they may be full-height, as at higher security locations. Turnstiles at security locations may be remotely operated or connected to ACCESS CONTROL SYSTEMS and used as MANTRAPS if necessary. Sensing devices for explosives and metal are often built into the frames of turnstiles.

twenty-four hour set detection (circuit) see PERMANENT CIRCUIT.

Type I, Type II, Type III a grading system comprising part of a standard developed by UL to define the extent of protection afforded by various levels of protection inside premises. See also Appendix 2.

UL Underwriter's Laboratories: a US-based testing and standards-issuing organisation funded by the insurance industry, with a particular interest in fire, security and safety equipment. An item approved by this body is said to be 'UL-listed'. In some locations such a listing may be mandatory for items such as fire extinguishers or smoke detectors.

ultra-lowlight (camera) see SIT.

ultrasonic motion detector a SENSOR used in VOLUMETRIC PROTECTION which detects motion. It operates by transmitting a high frequency sound (above human audibility range), which is reflected off objects within the protected area. If any change in these reflections takes place after the system is switched on, the frequency of the reflections changes (DOPPLER SHIFT), triggering an alarm.

ultrasonic sensor see ULTRASONIC MOTION DETECTOR.

ultraviolet light an electrical light source, usually portable, which makes visible anything marked with ultraviolet marking pens. An ultraviolet light is also used to verify 'invisible' signature strips in bank passbooks as well as making INVISIBLE STAINING COMPOUNDS visible.

Ultricon a closed-circuit television camera tube with operating characteristics similar to that of the NEWVICON but capable of producing satisfactory images only down to between 1 and 0.5 LUX.

undercover investigations the use of security or investigation personnel who are infiltrated into a plant or company in the guise of employees to carry out investigations into theft, sabotage, the use of drugs or similar matters.

underground protection system a term sometimes used to describe any BURIED DETECTION SYSTEM. Sometimes used as a proprietary name for a type of PRESSURE DIFFERENTIATING SYSTEM.

unit lock a complete MORTICE LOCK, normally operated by a PIN TUMBLER MECHANISM, supplied in a preassembled form ready for fitting. The preassembly and simplicity of fixing saves time and eliminates incorrect installation.

unset see SET.

upright lock see SASH LOCK.

UPS Uninterruptible Power Supply: a device or group of devices which will ensure a constant electrical supply for a limited period. Often used in computer installations to prevent loss of data during a period when mains (AC) power is subject to interruptions or fluctuations.

UV ULTRAVIOLET LIGHT.

V

vapour detection the principal method of explosive detection.

vapour detector see EXPLOSIVE DETECTOR.

vault a secure storage room, often located in a basement.

VDU Visual Display Unit: the screen or monitor of a computer unit or computer-controlled piece of equipment.

vehicle barrier a movable obstruction used to prevent free access to a site. Vehicle barriers are usually constructed in the form of a pivoting arm and may be motor operated or unpowered. Motor-operated devices (which may be electromechanical or electro-hydraulic) can be locally operated by a guard or can be interconnected with a CARD READER or other form of ACCESS CONTROL SYSTEM. Electrically operated gates could also be considered to be vehicle barriers. See also HYDRAULIC VEHICLE BARRIER.

vehicle immobiliser a device for rendering a vehicle unusable. Such devices can be mechanical (for example, a wheel clamp), electronic (in the form of a system which isolates the vehicle's fuel and ignition system), or a combination of the two.

vetting see BACKGROUND SCREENING, PV, NV, SECURITY CLEARANCE.

vibration detector see VIBRATION SENSOR.

vibration sensor a sensor which responds to specified levels of vibration. These may be INERTIA SENSOR based or piezoelectric-based, of a more sensitive type containing a 'tuning fork' which will respond only to a very narrow set of frequencies.

video motion detector see MOTION SENSOR (2).

video window the part of a closed-circuit television image selected by a MOTION SENSOR. The window is set up on the MONITOR and any change in the picture displayed in the window results in an alarm being triggered.

vidicon a conventional closed-circuit television camera tube which will provide an acceptable image at light levels down to 25 LUX. However, at this level of light, images can suffer from distortion due to movement. This is particularly so in the case of a scene viewed at night, when the headlights of vehicles will create severe 'blazing' on the monitor screen. See also NEWVICON.

visible staining compound a group of chemicals used to mark property and identify thieves. These chemicals are usually moisture-activated and when they come into contact with human skin, a reaction takes place producing a distinctive colour which is difficult to remove and will last for two to three days. See also EOSINE, MALACHITE GREEN, PARA-ROSALINE SULPHATE, RHODAMINE.

vision panel a small window set into a door. Such panels should not exceed 1.2 m square if the door is designated fire-resistant and must be made of wire-reinforced glass. See also GEORGIAN GLASS.

visitor badge a type of IDENTITY CARD or badge supplied to persons visiting a site on a temporary basis. Such cards or badges rarely contain a photograph of the visitor but are visibly different from the cards or badges worn by staff. Visitor identification should either be of the self-destructing type or should be issued in exchange for a driving licence or other document. See also PHOTO-BADGE EXCHANGE.

visual alarm a method of indicating that an ALARM CONDITION exists by means of a flashing light, STROBE LIGHT or other illuminated signal.

voice actuator a SENSOR which triggers an alarm when a human voice is detected. This type of sensor works by responding to the frequencies typical of human speech.

voice analysis a method of identification by BIOMETRIC SENSOR using the distinct and apparently unique characteristics of the human voice.

voice-grade multiplex signal MULTIPLEX SIGNALLING which is of an adequate quality to transmit speech.

voice-over observation the technique of combining a closed-circuit television system with equipment which enables an operator to talk to a person in the field of view. Such facilities are provided in DOOR VIEWER systems for domestic situations. The concept has also been used successfully at unmanned gates where an operator can question a visitor from a secure, remote location before deciding whether or not to operate a gate.

voice recognition system an ACCESS CONTROL technique based on VOICE ANALYSIS where digitised stored voice-samples are compared with spoken key phrases. Such systems have been used in very high security applications.

voice stress analyser see PSYCHOLOGICAL STRESS ANALYSER, POLYGRAPH.

volumetric detection see VOLUMETRIC PROTECTION.

volumetric protection the use of INFRA-RED, MICROWAVE or ULTRASONIC SENSORS to provide total protection for an entire room, space or building. In volumetric protection systems, the entire PROTECTED AREA is covered by one or more sensors. Also known as area or space protection. (See diagram 2 on p. 29)

volumetric sensor see VOLUMETRIC PROTECTION.

WAD World Association Of Detectives: a US-based association with international membership, set up to further the interests of those employed in the private investigation field.

Wafer see DISC TUMBLER.

walktest a method of testing the actual coverage of a VOLUMETRIC SENSOR. Using a device known as a walktest light connected to the CONTROL PANEL, it is possible to demonstrate the actual detection pattern limits and make any final adjustments to the system. Some sensors have an LED fitted to them which is always active and enables an occupier to know when a detector is functioning correctly. See also LATCHING.

walktest indicator see LATCHING.

walk under a method of defeating a VOLUMETRIC SENSOR by passing through the 'blind' area directly beneath the unit.

ward see WARDED LOCK.

warded lock a LEVER LOCK MECHANISM (usually of an older type) with internal wards, or obstructions, which block the entry or turning of an incorrect key. See also BULLET-WARDS.

warning output any signal, audible or visible which is intended to alert the KEYHOLDER (or other responsible person). Warning outputs serve to indicate a FAULT particularly in a SUPERVISED CIRCUIT or power supply. A warning output will also be sounded if a keyholder tries to switch on a system when one or more sensors are inoperable – for example, if a door or window has been left open.

watchman's clock a mechanical recording device used to ensure that a guard or watchman carries out regular patrols. Sometimes called a tour clock, the device consists of a clockwork mechanism which drives a paper tape or disk. Numbered keys are distributed along the patrol route and as each is inserted in the clock, a printing device imprints the time and the number of the key. The recording tape can only be accessed by a special key.

Recent advances in technology have made it possible for the device to be made much simpler so that it is now possible to buy a device the size of a pocket pager, known as an electronic guard patrol recording system.

This is carried by the guard and its base is inserted into sockets (or placed near readers) set at fixed points around the patrol route. The unit records the information from the sockets and at the end of a shift, the unit can be downloaded into a microcomputer which will print out a detailed log of the guard's activities.

watch tour a physical inspection or patrol of a location or building. In the US, this activity is sometimes called a guard tour.

Weigand card an ACCESS CARD using the properties of the WEIGAND EFFECT. (See table 1 on p. 17)

Weigand Effect the pulse-generating properties imparted to the ferromagnetic wires which are embedded inside certain types of ACCESS CARD. Some authorities spell this term 'Wiegand'.

weighbridge fraud a group of frauds which involve the misrepresentation of the weight of a vehicle and its load. Such frauds often involve collusion with weighbridge staff or can involve complex vehicle modification. Examples of such frauds range from the simple – increasing the apparent weight of a delivery by allowing the tarpaulin or tilt to fill with rainwater – to the complex. For example, one reported case indicates that a driver had arranged a system to pump mercury from a tank in his tractor to a reservoir in the semi-trailer. This of course enabled him to increase or decrease the weight of his load at will.

West Midlands Policy see BELL SHUTOFF.

wheel a type of DETAINER, manufactured in the shape of a small wheel.

Wheelbarrow a mechanised robot-like device used by EOD units in the UK and elsewhere. The device is tracked and can be fitted with a variety of accessories including a crane arm, shotgun or DISRUPTER, and an X-ray device. The Wheelbarrow enables EOD personnel to examine and deal with suspected IEDs with the minimum of exposure.

wicket gate a small (personnel) gate set in a larger one. Sometimes used to refer to an opening or window fitted with a grille, as in a bank or ticket office.

wide-gap contacts special door CONTACTS designed to be used in those applications where greater tolerances in respect of the distance between the REED SWITCH and its magnet are required. An example of this type of application would be for ROLLER SHUTTERS.

Wiegand Effect see WEIGAND EFFECT.

wind loading a measurement of the likely maximum wind speed an object is designed to withstand.

window bolt a SUPPLEMENTARY LOCKING DEVICE which can be morticed into a wooden window frame to provide unobtrusive security. It closely resembles a RACK BOLT but is shorter. (See diagram 13 on p. 97)

window foil see FOIL.

window stay a simple metal device used to keep a window in the open or partly open position. It has no security value unless fitted with a STAY LOCK.

window stay pin see LOCKING STAY PIN.

wipe-through card reader a CARD READER which reads the encoded data when an ACCESS CARD is simply passed through an open head rather than being inserted into a slot. The advantages of this type of unit are lower instances of mechanical damage and a faster throughput of users. This type of reader is also known as a swipe reader.

wired glass see GEORGIAN GLASS.

wired system see HARDWIRED.

wireless alarm (system) see RADIO ALARM SYSTEM.

X-ray system a security system which uses X-rays to inspect the contents of postal items, baggage and so on. Portable X-ray systems are also used by specialists dealing with IEDS.

Yodelarm a proprietary name for an electronic SOUNDER now frequently used in place of electric bells. A similar device is known as a 'Banshee'. These devices make a very loud, distinctive sound and are much more noticeable than other types of sounders. They are also to be preferred as they impose a lower current drain on a system than do bells.

Z

zone in the security context, a zone is defined as an area covered by a single alarm control circuit. Zones can be wired in series or in a loop and are connected directly to a CONTROL PANEL. Even the smallest alarm system must incorporate at least two zones to allow for the connection of PANIC BUTTONS to a 24-hour, or PERMANENT CIRCUIT which is always in ACTIVE MODE.

zone expander a device which permits more than one ZONE to be monitored using a single pair of wires.

zoom lens a single lens, often specified in closed-circuit television systems, which has a variable focal length, permitting the operator to obtain close-ups or distant views using a remote control.

Z

APPENDIX 1

British Standards Institute – Specifications Relating to Industrial Security

BS 1722 Fences

Part 1 Chain Link Fences
Supplement to Part 1 Gates and Gateposts for Chain Link Fences
Part 10 Anti-Intruder Chain Link Fences
Part 12 Steel Palisade Fences

BS 2740 Simple Smoke Alarms and Alarm Metering Devices

BS 3621 Thief-Resistant Locks

BS 4166 Automatic Intruder Alarm Termination Equipment in Police Stations

BS 4737 Intruder Alarm Systems in Buildings

Part 1 Requirements for systems with audible signalling only

Part 1 Section 1.1 Installation
Part 1 Section 1.2 Maintenance and records

Part 2 Requirements for systems with remote signalling

Part 2 Section 2.1 Installation
Part 2 Section 2.2 Maintenance and records

Part 3 Requirements for detection devices

Part 3 Section 3.1 Continuous wiring
Part 3 Section 3.2 Foil on glass
Part 3 Section 3.3 Protective switches
Part 3 Section 3.4 Radiowave Doppler detectors
Part 3 Section 3.5 Ultrasonic movement detectors
Part 3 Section 3.6 Acoustic detectors
Part 3 Section 3.7 Passive infra-red detectors
Part 3 Section 3.8 Volumetric capacitive detectors
Part 3 Section 3.9 Pressure mats
Part 3 Section 3.10 Vibration detectors

Part 3 Section 3.11 Rigid printed-circuit wiring
Part 3 Section 3.12 Beam interruption detectors
Part 3 Section 3.13 Capacitive proximity detectors
Part 3 Section 3.14 Deliberately operated devices

BS 5051 Security Glazing

Part 1 Bullet-resistant glazing for interior use
Part 2 Bullet-resistant glazing for exterior use

BS 5357 Code of Practice for the Installation of Security Glazing

BS 5544 Anti-Bandit Glazing (resistant to manual attack)

BS 5979 Direct Line Signalling Systems and Remote Centres for Intruder Alarm Systems

BS 8220 Security of Buildings Against Crime

Part 1 Dwellings

APPENDIX 2

Selected US Standards Relating to Industrial and Commercial Security

Fencing and Barbed Tape

ASTM Standards

> F 552-78 Definitions of Terms Relating to Chain Link Fencing
> F 567-78 Practice for Installation of Chain Link Fence
> F 626-79 Specification for Fence Fittings
> F 668-81 Specification for PVC-Coated Steel Chain Link Fence Fabric
> F 669-81 Specification for Strength Requirements of Metal Posts and Rails for Industrial Chain Link Fence

Chain Link Manufacturers Institute
Manual 2.13(d)

> Specification for Metallic-Coated Steel Chain Link Fence Fabric
> Standard Specification for PVC-Coated Steel Chain Link Fabric
> Industrial Steel Specification for Fence Rails, Posts, Gates and Accessories

Expanded Metal Manufacturers Association
Standards for Expanded Metal Mesh

US Military Specifications

> Barbed Tape Mil-B-52488
> Barbed Tape Concertina Mil-C-52489(MO)
> General Purpose Barbed Tape Obstacle Mil-B-52775A

Door Hardware

UL Standard
Key Locks UL 437

ANSI Standards

> Butts and Hinges A156.1 1970
> Locks and Door Trim A156.2 1975
> Exit Devices A156.3 1972

Door Control Closers A156.4 1972
Architectural Door Trim A156.6 1972
Template Hinge Dimensions A165.7 1972
Door Controls A156.8 1974

US Department of Justice, LEAA

Physical Security of Door Assemblies and Door Components
Physical Security of Sliding Glass Door Units
Physical Security of Window Assemblies

Lighting Standards

Illuminating Engineering Society

American National Standard Practice for Protective Lighting
American National Standard Practice for Roadway Lighting

Bullet- and Burglary-resisting Materials

UL Standard
Bullet-Resistant Materials UL 752
(Note: the Standard does not formally allocate 'Class Numbers' but these
are in general use).

Class 1 (Medium Power) Super .38 Automatic
Class 2 (High Power) .357 Magnum Revolver
Class 3 (Super Power) .44 Magnum Revolver
Class 4 (High Power) 30.06 Military Rifle

Burglary-Resistant Materials UL 972

Intruder Detection and Alarm Systems

UL Standard

Central Station Burglar Alarm Units and Systems UL 611
Robbery Systems UL 636
Installation, Classification and Certification of Burglar Alarm
Systems UL 681
Proprietary Burglar Alarm Units and Systems UL 1076

Federal Specification
Alarm Systems, Interior, Security W-A-00450B-(GSS-FSS)

NFPA Standard
Proprietary Signalling Systems NFPA 72-D

Guard Operations

NFPA Standards

Guard Services in Fire Loss Prevention NFPA 601
Guard Operations in Fire Loss Prevention NFPA 601A

Appendix 2

Cargo Security

US Customs Service Standard for Cargo Security
High Risk Cargo 19 CFR 4.30

Safes and Vaults

Safes

Insurance Services Office Standard

> Grade B (lowest)
> Grade C
> Grade E (highest)

UL Standards
These are rated according to the degree of resistance for a specified period against various types of attack and range from:

> TXTL60 (highest) tool- and explosive-resistive – 60 minutes
> TR30 (lowest) torch- and tool-resistive – 30 minutes

SMNA Standards
These are rated from UB1 (the highest, which equates with UL TXTL60) through B1 to R1 (the lowest, which has no equivalent UL listing).

Secure Storage of Documents

UL Standards
Insulated Record Containers

> Class 350 – 4 hour (A)
> Class 350 – 2 hour (B)
> Class 350 – 1 hour (C)

Insulated Filing Devices

> Class 350 –1 hour (D)
> Class 350 – 0.5 hour (E)

Fire-Resistant Safes

> Class 350 – 4 hour (A)
> Class 350 – 2 hour (B)
> Class 350 – 1 hour (C)

Vault Protection
Although bank vault equipment is tested and labelled by UL, the vaults themselves are not.

Insurance Services Office
Bank vaults are rated from Class 1 through 4, then 5R, 6R, 9R, and 10R. Classes 1 to 4 are directly equivalent to Classes B to E for Mercantile Vaults.

Bank Protection Act requirements are satisfied by Classes 5R to 10R provided that the vault floors, walls and ceiling are constructed of concrete at least 12″ thick.

An additional range of classes 11 to 13 are not listed by the ISO but are a consensus standard reached between the vault industry and banking for vaults which require protection in excess of the 10R standard.

SELECT BIBLIOGRAPHY

Alth, Max, *All About Locks and Locksmithing*, New York, Hawthorn Books, 1972.

Anderson, E. E., *Bank Security*, Woburn, Mass., Butterworth, 1981.

Astor, Saul D., *Loss Prevention: Controls and Concepts*, Los Angeles, Security World Publishing, 1974.

Balnard, R. L., *Intrusion Detection Systems*, Woburn, Mass., Butterworth, nd.

Berger, D. L., *Security for Small Businessmen*, Woburn, Mass., Butterworth, nd.

Bilek, Klotter, Keegan and Federal, *Legal Aspects of Private Security*, USA Anderson Publishing Co, 1981.

*Bose, Keith, *Video Security Systems*, Woburn, Mass. Butterworth, 1982.

Buzby, Walter J. and Paine, David, *Hotel and Motel Security Management*, Los Angeles, Security World Publishing, 1976.

Byrne, Dennis E. and Jones J., *Retail Security – A Management Function*, Leatherhead, 20th Century Security Education, 1971.

Carrol, John M., *Computer Security*, Los Angeles, Security World Publishing, nd.

Carson, Charles R., *Managing Employee Honesty*, Los Angeles, Security World Publishing, 1977.

*Clarke, R. V. G. and Mayhew, P. (eds) *Designing Out Crime*, London, Home Office, 1980.

*Clutterbuck, Richard, *Kidnap and Ransom – the Response*, London, Faber, 1978.

Colling, Russel L., *Hospital Security*, Los Angeles, Security World Publishing, nd.

*Comer, Michael, *Corporate Fraud*, London, McGraw-Hill, 1980.

Comer, Michael (ed.), *Find The Truth*, London, Network Security Management Limited, 1984.

Cunningham, John E., *Security Electronics*, USA, Howard W. Samms & Co, 1973.

Currer-Briggs, Noel (ed.), *Security Attitudes and Techniques for Management*, London, Hutchison, 1968.

*Currer-Briggs, Noel and Hamilton, Peter, *Handbook of Security*, London, Kluwer Publishing, 1974 (looseleaf with quarterly updates).

Dorey, Frederick, *Aviation Security*, London, Granada, 1983.

Earnshaw, Len, *Construction Site Security*, London, Construction Press, and New York, Longman, 1984.

Essentials of Security Lighting, London, The Electricity Council, 1981.

Fennelly, Lawrence J., (ed.) *Handbook of Crime and Loss Prevention*, Woburn, Mass. and London, Butterworth, 1982.

Fine, L. H., *Computer Security*, London, Heinemann, 1983.

Finneran, E., *Security Supervision*, Woburn, Mass., Butterworth, nd.

Green, G. and Farber, R. C., *Introduction to Security*, Los Angeles, Security World Publishing, nd.

*Hamilton, Peter, *Espionage, Terrorism and Subversion in an Industrial Society*, Leatherhead, Peter A. Heims Limited, 1978.

Hasler, Gordon, *Protect Your Property and Defend Yourself*, London, Penguin Books, 1982.

Healy, Richard J., *Design for Security*, New York, John Wiley and Sons, 1968.

Healy, Richard J. and Walsh, Timothy J., *Industrial Security Management – a Cost-Effective Approach*, New York, American Management Association, 1971.

Heims, Peter, *Countering Industrial Espionage*, Leatherhead, 20th Century Education, 1982.

Hemphill, Charles F. Jr., *Modern Security Methods*, Englewood Cliffs, NJ, Prentice Hall, 1979.

Hopf, Peter S. (ed.), *Handbook of Building Security, Planning and Design*, New York, McGraw-Hill, 1979.

Hughes, Denis, *Contract Security*, Leatherhead, 20th Century Security Education Ltd, 1982.

Hughes, Denis, *Security Officer's Handbook*, Leatherhead, 20th Century Education Ltd, 1980.

Hughes, Denis and Bowler, P., *The Security Survey*, Aldershot, Gower Publishing, 1982.

Hughes, Mary M. (ed.), *Successful Retail Security*, Woburn, Mass. and London, Butterworth, nd.

Jones, Peter and Byrne, Dennis, *Store Detective's Handbook*, Leatherhead, Peter A. Heims Limited, 1984.

Kingsbury, Arthur A., *Security Administration, an Introduction*, Springfield, Ill., Charles C. Thomas, 1973.

Knowled, Graham, *Bomb Security Guide*, Los Angeles, Security World Publishing, 1976.

Luis, Ed San, *Office and Office Building Security*, Los Angeles, Security World Publishing, 1975.

McKnight, Gerald, *Computer Crime*, London, Michael Joseph, 1973.

Merrigan, Guy and Wanat, *Forms for Safety and Security Management*, Woburn, Mass., Butterworth, 1981.

Moore, Kenneth C., *Airports, Aircraft and Airline Security*, Los Angeles, Security World Publishing, nd.

Momboisse, Raymond M., *Industrial Security for Strikes, Riots and Disasters*, Springfield, Ill., Charles C. Thomas, 1968.

Morneau, R. H. Jr and Morneau, G. E., *Security Administration*, Woburn, Mass., Butterworth, 1982.

Naidich, Arnold, *Protect Your Company from A–Z*, New York, Man and Manager Inc., 1976 (looseleaf).

Neill, W. J., *Modern Retail Risk Management*, Woburn, Mass., Butterworth, nd.

Newman, Oscar, *Architectural Design for Crime Prevention*, Washington DC, US Government Printing Office, 1971.

*Newman, Oscar, *Defensible Space*, New York and London, Macmillan, 1972, and Architectural Press, 1973.

*Oliver, Eric and Wilson, John, *Practical Security in Commerce and Industry*, Aldershot, Gower Publishing, 1978.

Oliver, Eric and Wilson, John, *Security Manual*, Aldershot, Gower Publishing, 1979.

O'Toole, George, *The Private Sector*, New York, W. W. Norton & Co, 1978.

Pascall, A. M., *Hospital Safety and Security*, USA, Aspen Publications, 1977.

Powell, W. T., *Campus Security and Law Enforcement*, Woburn, Mass., Butterworth, 1981.

Pyner, Barry, *Design Against Crime*, London, Butterworth, London, 1983.

Ricks, Tillet and Van Metel, *Principles of Security*, USA, Anderson Publishing Co, 1982.

Rockley, L. E. and Will, D. A., *Security – its Management and Control*, USA, Business Books, 1980, and London, Hutchison, 1981.

Rogerson, David, *Make Your Home Secure*, Newton Abbott, David & Charles, 1984.

Schweitzer, J. A., *Managing Information Security*, Woburn, Mass., Butterworth, nd.

*Sennewald, Charles A., *Effective Security Management*, Los Angeles, Security World Publishing, 1978.

Sennewald, Charles A., *The Process of Investigation*, Woburn, Mass., Butterworth, nd.

*Sykes, J. (ed.), *Designing Against Vandalism*, London, The Design Council, 1979.

Trimmer, W. W., *Understanding and Servicing Alarm Systems*, Woburn, Mass., Butterworth, 1981.

Underwood, F., *The Security of Buildings*, London, Architectural Press, 1984.

Walker, P., *Electronic Security Systems*, Woburn, Mass., Butterworth, nd.

Walsh, Dermot, *Break-ins – Burglary from Private Houses*, London, Constable, 1980.

Walsh, Dermot, *Shoplifting – Controlling a Major Crime*, London, Macmillan, 1978.

Walsh, Timothy J. and Healy, Richard J., *Protection of Assets Manual*, Sacramento, The Merritt Co, 1974, (looseleaf with monthly updates).

Warne, Albert and Brown, Don, *Industrial Security*, Chichester, Barry Rose Publishing, 1975.

Weber, Thad L., *Alarm Systems and Theft Prevention*, Los Angeles, Security World Publishing, 1973.

Whiteside, Thomas, *Computer Capers*, New York, Thomas Y. Crowell, 1978, and London, Sidgwick & Jackson, 1979.

Wright, K. G., *Cost-Effective Security*, London, McGraw-Hill, 1982.

Yallop, H. J., *Protection Against Terrorism*, Chichester, Barry Rose Publishing, 1980.